The Future of
Lean Sigma Thinking
in a Changing
Business Environment

THE FUTURE OF LEAN SIGMA THINKING IN A CHANGING BUSINESS ENVIRONMENT

DAVID ROGERS

CRC Press
Taylor & Francis Group
Boca Raton   London   New York

CRC Press is an imprint of the
Taylor & Francis Group, an **informa** business
A PRODUCTIVITY PRESS BOOK

CRC Press
Taylor & Francis Group
6000 Broken Sound Parkway NW, Suite 300
Boca Raton, FL 33487-2742

Printed in the United States of America on acid-free paper
Version Date: 20110427

International Standard Book Number: 978-1-4398-5102-9 (Hardback)

Visit the Taylor & Francis Web site at
http://www.taylorandfrancis.com

and the CRC Press Web site at
http://www.crcpress.com

Dedicated to my

mother,

Vera Rogers

Contents

Preface

As a broad-brush statement, the world has reached a turning point. The effects of decades of neglecting our effect on the environment is coming back to haunt us. Conversely, the consequences of neglecting the effects of toxic debt are now a major concern to everyone as the ripple effect gathers pace and will ultimately reach every corner of the planet. By way of an example, reduced aid and rising sea levels are causing devastation to coastal farmers eking out a living, particularly in archipelago countries.

So how can *we* in the business sector solve these problems? Well, of course, the simple answer is that we can't solve them on our own. What we can do is use the next year or two to strengthen our businesses so that as the global recession recedes, for it surely will at some point in the not-too-distant future, our businesses are healthier and stronger, ready to absorb as many orders as we can process so that we maximize the economic upturn.

Many business tools are used to reduce waste and maximize potential. The continuous improvement process known as *lean thinking* is often the method of choice, well worth the investment in time and effort. Lean thinking has been with us now for a few years and significantly added to the bottom line of all those companies large and small that have implemented its techniques and adopted its philosophies.

However, some might argue that the various organizations training change agents in lean thinking techniques need to reevaluate the "process" of process improvement to determine how, if at all, lean thinking should be modified to cope with the change in the current business climate. For this loose conglomerate of disparate training companies almost appears to have a "corporate" culture of its own. And with no obvious hierarchy for introducing new ideas into the process of lean thinking implementations, these very concepts might be holding back progress.

I must state that I am a fan of lean thinking. But I also hold these thoughts in my mind:

Of the top 100 companies from 1900, only 16 exist today.

Fortune magazine published its first list of the world's 100 largest companies in 1956—today, only 29 of those companies remain on the list and the number is falling!

During the 1980s, 230 companies (46%) disappeared from the Fortune 500 list.

Size and reputation do not guarantee success.

This is an old statement. Eastman Kodak and Polaroid, among others, may add to the endangered or critical list if they are not careful. Some if not all of these large corporations lost their way because they failed to appreciate the changes taking place, or were too big to react fast enough to changing market conditions.

In many ways, organizations that implement lean thinking concepts might suffer the same fate. An update in lean thinking now might help us all through these difficult times, making industry stronger while having a less detrimental impact on raw materials and the wider environment.

There may be no better example on which to base this study than the global motor industry, for most of the principles of lean thinking originated in the Toyota Motor Company. It seems like every day, the press report of consolidation, redundancies, or government bailout issues within that industry. Painful though it is now, it is also the time to consider how these companies will look postrecession, for then will be an equally testing time since vehicles wanted in 2010 may not meet the specifications of cars demanded in 2011 or beyond....

Acknowledgments

Writing this book was interesting from many perspectives. I received my Six Sigma training in the early 1990s and coordinated all further Six Sigma training that took place in the research and development laboratories at Kodak Limited. I went on to work on a large project to implement design for manufacturability concepts in the manufacturing division, a stint implementing process verification technology across the site, prior to moving to a leadership position as head of the process research and development group. Along the way, I undertook management blackbelt training and subsequently trained in and then implemented Toyota Production System (TPS) concepts, which at Kodak Limited were rebranded to suit the culture of the company. In addition, I held the post of visiting professor in the Business Psychology Centre at the University of Westminster. Indeed, I am still a visiting lecturer at that institute.

Nowadays, TPS concepts are often referred to as *lean thinking*, although some companies in other markets have also rebranded the concepts and provided themselves with an in-house name and feel. Yet further companies have called the concepts *lean* or *Six Sigma*, and some further companies have simply trained their blackbelt personnel in TPS concepts. Occasionally, one also reads the term *just in time*.

Yet with all of this interest across many manufacturing companies, there has been very little penetration in the chemical industry compared with the potential number of chemical companies. This particular industry has, in part, been implementing concepts known as *process intensification*, a methodology which is also of interest in terms of process improvements as these techniques also offer a "green" flavor.

From my own perspective and having been involved in quality concepts (as a role within my other duties) for the last fifteen years or so, it is perhaps timely to put pen to paper and draw on my own experiences to help companies thinking of becoming involved. These concepts provide a collection of ideas, references, and experiences that should help to shortcut the inevitable pain and heartache these implementations can invoke.

There are many people I would like to thank for training me or for their time in helping me to implement some of the concepts. They are, in no particular order (or seniority!), Bob Holwill, Terry Wright, John Depoy, Paul Campion, Brendan Hederman, Duncan MacIntosh, Don Wild, Dave Balicki, Bob Hopkins, Fred Hamaker, Bill Slater, Mary Farmer, Paul Osborn, Keith Hall, James Christie, Ken Sankey, James Treadaway, Jon Knight, John Beresford, John Marr, James Dauncey, Andy Phillpott, Cem Ural, Martin Arnold, Dave Raeburn, Morris Chung, Roger Houghton, Jeremy Foster, Roger Winter, and Peter Charnley. Thazi Wisudha-Edwards provided some timely information concerning process intensification; thank you, Thazi.

Compiling lists is always fraught with problems as there is inevitably someone who deserves to be mentioned but who through my fault has been neglected. I apologize to anyone who finds themselves in that category. In preparing a manuscript such as this, I also would like to thank my wife, Carolyn, for proofreading and providing endless hours of support and our two sons, Adam and James, who would have liked more of my time now and again.

1

Quality Initiatives

Perhaps before launching into deep and meaningful discussions regarding some of the quality improvement methods, we should spend a few moments reflecting on what is meant by the term *quality*. Dictionary definitions use phrases like "degree of excellence" or "relative nature or kind." Certainly, the quality of product A can be measured relative to the quality of product B fairly easily, providing of course that the two products perform a similar function.

Genichi Taguchi formulated a method of relating cost to customer satisfaction (which also holds valid for product specifications), shown in Figure 1.1 and often referred to as the *Taguchi loss function*. This relates customer dissatisfaction to loss of income from the product under discussion. The further the product is from customers' expectations, the more dissatisfied customers become. One would like to think that the marketing and sales force have correctly interpreted customer needs, which the research and development communities have translated into a product, which has been manufactured within the best tolerances available to them. So manufacturing any product on aim exactly meets customer expectation, which results in maximum sales and therefore profits for the company. A deviation from the target toward the upper or lower specification may therefore result in loss of sales. Of course, the caveat here is that there may be no alternative product available and so a customer might buy a batch of products knowing that the batch contains variable products simply because there is no alternative.

Customer expectations for the quality of a given product often change depending on the price paid for that item. Put another way, you might tolerate a lower standard of quality for an item that costs only a few dollars. Conversely, you would want a high-quality item when paying several hundred dollars.

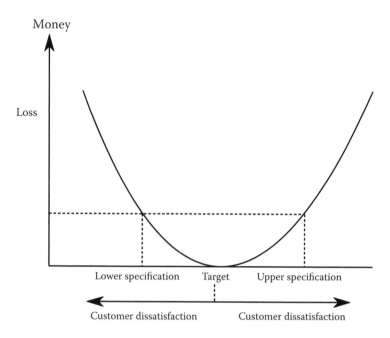

FIGURE 1.1
The Taguchi loss function.

In practice, we would therefore expect to see a series of curves for the Taguchi loss function where each curve is price related (Figure 1.2). These curves (Figure 1.2) show the concept graphically in that as price increases, quality expectations also rise, all other things being equal. It is certainly the case in some organizations where the more discerning customers are often sent product that is "cherrypicked" (i.e., the product is sorted prior to dispatch so it only arrives within specification).

Furthermore, there are some brave engineers who have related process capability with loss in revenue using the same thinking as used in the Taguchi curves (and borrowing from the Six Sigma methodology). This modified plot relates a *process capability index* (Cp or Cpk) in the way shown in Figure 1.3. Overlaying some measure of process capability over a Taguchi loss function seems like it might be really useful. All you have to do is to get the engineers to fix the process and you are home free, with a perfect product every time! Actually, it is much more complex as someone must determine key process variables, set aims and limits, draft corrective action guidelines, and more. Let's have a look at a simple process for establishing key process variables used in at least one major multinational

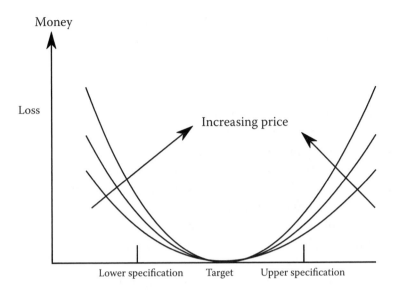

FIGURE 1.2
Customer expectation price curves.

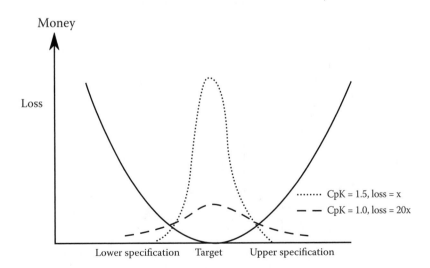

FIGURE 1.3
The Taguchi loss function versus process capability.

corporation. This somewhat small diversion is shown here to demonstrate the complexity of the undertaking. Still, nothing ventured, nothing gained.

The basic questions listed below need to be addressed:

- What are the steps in your process?
- What could go wrong with the process?
- How could you monitor potential process problems?
- Will more than one variable need to be monitored?
- Can you interface the control system to a data logger?
- Is there a technology gap in sensors, control, or the like?
- How will the variables be monitored?
- What happens when a variable goes into alarm?

A diagram such as that outlined in Figure 1.4 can then be used to structure the potential verifiers. Figure 1.4 shows the higher order function on the left-hand side of the diagram, in other words, the activity that you are trying to achieve. It might be a function such as assemble parts. The first column to the right would then list all of the activities that have to be achieved in order to perform the higher function. One then asks the "how" question of each of these new functions, and so on. Eventually, at the right-hand side of the diagram, one has all of the control parameters

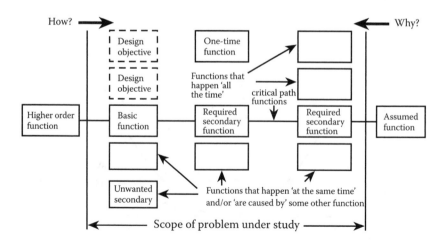

FIGURE 1.4
Relationship between higher order and assumed functions.

that are needed to undertake the function described on the left-hand side of the diagram.

Actually, this list of assumed functions is also a list of potential problems or barriers to a successful manufacturing operation. The reasoning here is that if all of these assumed functions perform at their optimum, then the higher order function will be successfully undertaken. A diagram such as that outlined in Figure 1.5 can then be used to look for potential verifiers for the assumed functions.

Using the methodology described above, there is often more than one potential problem. All of them should be listed on the left-hand side of Figure 1.5 as potential problems. One then assesses failure impact and failure frequency, and generates a list of potential verifiers for that potential problem.

Some companies then use numbers to replace the terms *high, medium,* and *low* and create a ranked list of verifiers. This exercise may determine that there is a technology gap for a particular verifier. Once the list has been generated, there are two scenarios (see Figure 1.6). This methodology has proven to be successful for some companies, but may be more complex than is needed for others. The complexity of the system used to determine

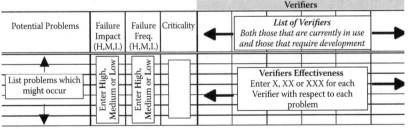

Process Zone :	For which Process Zone does this apply?
Functional Block :	What is the Functional Block?
Target :	What is the Target (Objective) of this Functional Block?
Goal :	What goal does this impact?
Control Strategy :	What control strategies are currently is use?
Alt. Control Strategy :	Are there alternate strategies?
Gap :	What Gaps currently exist?

FIGURE 1.5
A typical verifier table.

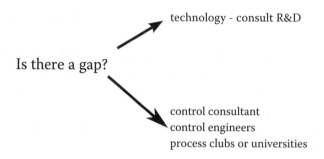

FIGURE 1.6
Technology gap choices.

and implement process verifiers is entirely dependent on the types of products that are manufactured and the complexity of the process.

So there are issues in product design and the need to ensure that product yield is as high as possible during the various steps. There is the effect of the starting component variability as the new product works its way through the manufacturing scale-up operation. This is but one scenario, the adverse effects of which can be reduced if the design team has production experience. The more skilled research and development communities are much more attuned to the possible downstream pitfalls and can help to mitigate the effects. It is perhaps surprising that only a handful of research and development scientists spend much time in their respective manufacturing divisions as product technologists, which is certainly the case in larger companies.

The issue of process-induced defects is a major concern to any industry, including software. It is just as likely that there will be a software bug in a program as there is to be an issue with a manufacturing process-induced defect, for example. But what level of control is appropriate? Section 7.2 in Chapter 7 discusses the whole issue of process verification. It is sufficient to mention here that setting up an automated system is complex and not appropriate for all processes and businesses, particularly where margins are small because the systems can be expensive to set up and run.

Understanding process-induced defects is specific to a particular aspect of the manufacturing process. There are some quality processes that look at the problem in a much more holistic way. Before discussing some of them in more detail, however, it will be instructive to look at some historical "movers and shakers," just to place a few quality processes in context. Two of the most influential "quality gurus" who worked across the world, particularly after World War II, were W. Edwards Deming and Joseph Juran. Let's have a look at just a small sample of their extremely influential work in the immediate postwar years.

1.1 DEMING AND JURAN

Deming was perhaps the first person to recognize that quality issues were solved only if two fundamental problems were addressed, namely, management and the process of change. Arguably his fourteen points of management are still relevant today. In the preface to his book *Out of the Crisis* (Deming, 1986), Deming made the point that

> it is no longer socially acceptable to dump employees on the heap of unemployment. Loss of market, and resulting unemployment, are not fordained. They are not inevitable. They are manmade. The basic cause of sickness in American industry and resulting unemployment is failure of top management to manage. He that sells not can buy not.
>
> The causes usually cited for failure of a company are costs of start-up, over-runs on costs, depreciation of excess inventory, competition—anything but the actual cause, pure and simple bad management. (p. ix)

It has a certain resonance, even during the current global financial meltdown. Just for reference, his fourteen points of management were as follows:

- Create constancy of purpose for improvement of product and service.
- Adopt the new philosophy.
- Cease dependence on inspection to achieve quality.
- End the practice of awarding business on the basis of price tag alone. Instead, minimize total cost by working with a single supplier.
- Improve constantly and forever every process for planning, production, and service.
- Institute training on the job.
- Adopt and institute leadership.
- Drive out fear.
- Break down barriers between staff areas.
- Eliminate slogans, exhortations, and targets for the workforce.

- Eliminate numerical quotas for the workforce and numerical goals for management.
- Remove barriers that rob people of pride of workmanship. Eliminate the annual rating or merit system.
- Institute a vigorous program of education and self-improvement for everyone.
- Put everybody in the company to work to accomplish the transformation. (Deming, 1986, p. 23)

Deming observed the Japanese started to realize that improvements in quality led to a corresponding improvement in productivity during 1948 and 1949; that's a long time ago! His observations of the Japanese were often made firsthand as Deming was a frequent visitor to Japan in the late 1940s, the 1950s, and beyond, during which he acted as a consultant. This was the postwar era of the rebirth of Japanese manufacturing, and many of Deming's ideas formulated or developed in Japan were implemented around the world. There are certainly many references to Deming's work, as we shall see later.

Of course, there are some pitfalls with management panaceas. There is the concept that measures of productivity lead to improvements in productivity. They don't, or at least not on their own. Just because you have the data doesn't mean that there is the technology to make improvements or indeed that things stand still. I have been a school governor for thirteen years or so and have been part of the process of setting targets for youngsters who are eight years old. Of course, each year group has a mixed ability: some children develop later than others, some arrive from war-torn parts of the world as refugees, and yet other children have special needs of a different nature from those of the previous year group. The mix of issues changes on a yearly basis. Yet the mandated targets continue to rise without those in central government understanding the issues at the grassroots.

Deming also proposed the "deadly diseases," which were as follows:

- Lack of constancy of purpose to plan product and service that will have a market, keep the company in business, and provide jobs
- Emphasis on short-term profits: short-term thinking (just the opposite from constancy of purpose to stay in business), fed by fear of unfriendly takeover, and by push from bankers and owners for dividends

- Evaluation of performance, merit rating, or annual review
- Mobility of management; job hopping
- Management by use of visible figures, with little or no consideration of figures that are unknown or unknowable
- Excessive medical costs—peculiar to the United States (and beyond the scope of this book) (Deming, 1986, p. 98)

I find these diseases and obstacles interesting in that Deming formulated his ideas many years before the credit crunch which started with defaults on bad loans—so called subprime or toxic debt—in 2008. There are also many corporations who have lost their way in constancy of purpose such as Eastman Kodak and Polaroid. Both of these corporations knew that the future of photography was almost certain to be related to digital photography, yet both prevaricated for such a long time that they lost market share at a time when the photospace market was growing!

Perhaps of equal importance for the Japanese from Deming's various visits was the cycle shown in Figure 1.7. In his book *Out of the Crisis*, Deming commented,

> The perception of the cycle **Plan Do Check Action** came from Walter A. Shewhart, Statistical Method from the Viewpoint of Quality Control (Graduate School, Department of Agriculture, Washington, 1939; Dover, 1986) p 45. I called it in Japan in 1950 and onward as the Shewhart cycle. It went into immediate use in Japan under the name of the Deming cycle, and so it has been called ever since. (p. 88 n. 9)

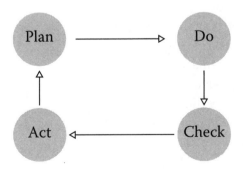

FIGURE 1.7
The plan, do, check, action (PDCA) cycle.

FIGURE 1.8
The Juran trilogy.

This cycle continues to be credited to Deming, despite his attempt to correct the situation. Of course, Deming was one of a few consultants who worked at this time in Japan. Joseph Juran also paid many visits to the country.

Arguably, Joseph Juran had as much influence in Japan as did Deming, during a similar time frame. Born on 24 December 1904, Juran died on 28 February 2008 at the age of 103. He continued to develop the Juran management system and develop the Juran Institute[1] for most of his adult life and was publishing books up until 2003 (*Architect of Quality*).

Much of his early thinking was distilled into a cycle which has become known as the Juran trilogy (see Figure 1.8). Juran promoted the concept that quality planning should start at product development:

- Product development.
  - Meet the needs of the customers.
  - Meet the needs of the suppliers.
  - Be competitive.
  - Optimize the combined costs of the customers and suppliers.
- Optimize product design.
- Process development.
  - Review product goals.
  - Operating realities.
  - Carryover of prior process designs.
  - Process control.
- Transfer to operations.
  - Proof of process capability.
  - Proof of process controllability.

- Transfer of know-how.

Juran not only suggested how to build quality into a product but also suggested how to build the organization structure:

- Identify the operations which need to be performed.
- Assign responsibility for doing these operations, whether internal or external agencies.
- Divide the total work pile into logical parcels of work, called *jobs*. In general, a job consists of a collection of one or more operations, so chosen that it is feasible to recruit or train people to carry out that collection.
- Define the responsibilities and authorities associated with each job.
- Define the relationships of each job to other jobs. These relationships include the following:
  - The hierarchical relationship (chain of command)
  - The communication and coordination patterns through which interdepartmental activities join to carry out specific purposes
- Orchestrate the work of the internal and external agencies so that the company's mission is carried out in an optimal manner.

Of course, both Deming and Juran provided solutions to many quality problems, from management best practices to detailed implementations. The above brief descriptions of their work, concentrating more on management and infrastructure, is complemented by the brief description of the quality systems of Six Sigma and the Toyota Production System.

1.2 SIX SIGMA

In the 1970s, some American companies experienced a wake-up call. In his paper "The Six Sigma Revolution," Pyzdek tells the following story:[2]

When a Japanese firm took over a Motorola factory that managed Quasar television sets in the United States in the 1970's, they promptly set about making changes in the way the factory operated. Under Japanese management, the factory was soon producing TV sets with 1/20th the number of defects they had produced under Motorola management. They did this using the same workforce, technology and designs, making it clear that the problem was Motorola's management. Eventually, even Motorola's own executives, in this case Art Sundry, had to admit 'our quality stinks'.... Finally, in the mid 1980s, Motorola decided to take quality seriously. Motorola's CEO at the time, Bob Galvin, started the company on the quality path known as Six Sigma and became a business icon largely as a result of what he accomplished in quality at Motorola. Today, Motorola is known worldwide as a quality leader and a profit leader. After, Motorola won the Malcolm Baldrige National Quality Award.[3]

Some years earlier, a senior staff engineer working at Motorola's Government Electronics Group (GEG) called Mikel Harry created a roadmap for improving costs at GEG by improving product design and reducing production time. Under Harry's leadership, a group of engineers within GEG demonstrated the potential of their thinking, culminating in a paper titled "The Strategic Vision for Accelerating Six Sigma within Motorola." Bob Galvin realized that this work merited further investment, leading to the implementation of Six Sigma across a wider area of the company than Harry could have reached on his own. In 1990, Gavin asked Harry to form Motorola's Six Sigma Research Institute at Schaumberg, Illinois. Other companies that participated in the institute in the early days were IBM, Texas Instruments Defence Group, Digital Electronics, Asea Brown Boveri, and Eastman Kodak.

While Mikel Harry is credited with the birth of the Six Sigma revolution at Motorola, it was a colleague of Harry's at Motorola, Bill Smith, who first studied the correlation between the number of repairs needed for a product during its useful life and how often the product had been repaired or reworked during the manufacturing process.[4] Smith came to the conclusion that there were bound to be some defects that were missed during the manufacturing process and so were not put right prior to the product sale. On the other hand, if a product was made with no defects, then the product rarely failed. Harry continued,

[W]as the effort to achieve quality really dependent on detecting and fixing defects? Or could quality be achieved by preventing defects in the first place through manufacturing controls and product design....
But What Exactly Is Six Sigma??? (Harry and Schroeder, 2006, p. viii)

Harry is often quoted as describing Six Sigma as "first a statistical measurement, second a business strategy and third a philosophy" (Harry and Schroeder, 2006, condensed from pp. 6–8).

But what does that mean? Pyzdek commented,

> Six Sigma is a rigorous, focused and highly effective implementation of proven quality principles and techniques. Incorporating elements from the work of many quality pioneers, Six Sigma aims for virtually error free business performance. Sigma, s, is a letter in the Greek alphabet used by statisticians to measure the variability in any process. A company's performance is measured by the sigma level of their business processes. Traditionally companies accepted three or four sigma performance levels as the norm, despite the fact that these processes created between 6,200 and 67,000 problems per million opportunities! The Six Sigma standard of 3.4 problems per million opportunities is a response to the increasing expectations of customers and the increased complexity of modern products and processes.
>
> If you're looking for new techniques, don't bother. Six Sigma's magic isn't in statistical or high-tech razzle-dazzle. Six Sigma relies on tried and true methods that have been around for decades. In fact, Six Sigma discards a great deal of the complexity that characterized **Total Quality Management (TQM)**. By one expert's count, there were over 400 TQM tools and techniques. Six Sigma takes a handful of proven methods and trains a small cadre of in-house technical leaders, known as Six Sigma Blackbelts,[5] to a high level of proficiency in the application of these techniques. To be sure, some of the methods used by blackbelts are highly advanced, including the use of up-to-date computer technology. But the tools are applied within a simple performance improvement model known as DMAIC, or Define-Measure-Analyse-Improve-Control. (Pyzdek, 2003[6])

Harry is fond of using the following analogy when describing the difference between Six Sigma and 3–4 Sigma to executives that visit his Six Sigma Academy:

> Each person sitting in the classroom is there because the airline's record in getting passengers safely from one city to another exceeds Six Sigma, with less than one-half failure per million. However, for those whose bags did not arrive with them, it's because the airline's baggage operations are in the 6,000–23,000 defects per million or 3.5–4 sigma. (Harry and Schroeder, 2006, p. 14)

For many years, the University of Motorola employed consultants to train various industry sectors on how to apply Six Sigma. Motorola published the following steps in their training literature:

The Motorola Six Steps to Six Sigma

- Identify the physical and functional requirements to the end product which are necessary to satisfy the requirements of:
  - customer's intended use of the product
  - foreseeable misuse of the product
  - environment in which the product is used
  - end life disposal of the product
  - regulatory agencies and applicable standards
- Identify the characteristics of the components and assemblies of the product which are key to meeting the end-product requirements
- For each key characteristic, determine whether it is controlled by the piece part, the assembly process, or a combination of both
- For each key characteristic, determine the target value which minimizes the effects of variation in that characteristic on the successful outcome, and determine the maximum allowable range of that characteristic which can be tolerated by the design
- For each key characteristic, determine the variation which can be expected in that characteristic based on the known capability of the process selected to reproduce it
- If the $Cp < 2$, and the process variation is benchmarked as best in class, then seek an alternative product design which eliminates the need, or increases the allowable variation for that characteristic
  - If the $Cp < 2$ and the process variation is not best in class, then seek an alternative process design which will result in acceptable variation within the allowable range
- If $Cp > 2$ but $Cp < 1.5$, then seek an alternative process design which will result in proper centering of the characteristics)

Six Sigma training discusses the issues of formulations under the topic of *design for manufacturability* (DFM). This catchall phrase encompasses such issues as batch-to-batch variability of starting materials through intended use and misuse of product to eventual disposal.

Overrun	Activity	Lost Revenue (%)
30%	Cost overrun	2
10%	Production cost increase	4
10%	Price reduction	15
6 months	Late delivery	33

FIGURE 1.9
The relationship between overrun costs and lost revenue.

There are many issues that need to be addressed in product design. Standard custom and practice is to use a phases and gates system which provides opportunities to assess the design through critical stages of the development process. Figure 1.9 provides some averaged data which compare cost overruns with late delivery. These help the research, development, manufacturing, marketing, and sales project team members to stay focused.

The new product *must* be available on time and defect-free; it is of paramount importance for most supply chain partners as well as from the cost perspective. There are many pitfalls to DFM. Perhaps the issue of supplier variability mentioned previously is noteworthy. Figure 1.10 shows two curves, one of which represents the starting component's variability used to design the initial product. It may also be the case that these batches of starting components were used in the pilot scale-up experiments. It is sometimes the case, however, that the large number of batches of starting materials needed to make production quantities of product results in the variability reflected by the larger of the two curves in Figure 1.10. Under

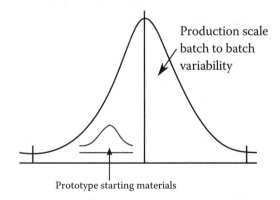

Production scale batch to batch variability

Prototype starting materials

FIGURE 1.10
A potential drawback to using small quantities of materials during prototype design.

these circumstances, the design of the prototype may not be sufficiently robust to successfully manufacture the desired product using the production quantities of starting materials.

1.3 THE TOYOTA PRODUCTION SYSTEM

For many years, it appeared as if Western quality initiatives revolved around defect reduction through efforts such as measuring indirect work (Whitmore, 1971), Six Sigma, and latterly New Six Sigma, whereas the Eastern quality thrust was more concerned with product and process design which eliminated defects from being produced at all. The most comprehensive methodology for defect elimination through a blend of a number of initiatives was developed at the Toyota Motor Company. This quality ethos or methodology is known as the *Toyota Production System* (TPS). Most of the elements of TPS are now taught during blackbelt courses along with Six Sigma best practices. These concepts took on a life of their own as *lean manufacturing* or *lean thinking*, mainly through the work of Womack and Jones.

Such is the importance of TPS, or lean thinking, that Chapter 3 has been devoted to discussing the origins and current methodology of TPS. That aside, there is a need to briefly mention here that TPS is actually a series of initiatives that were implemented within the Toyota Motor Company over many years. Figure 1.11 shows that work started on the first of these modules as the Japanese industry began its rebuild program in the mid- to late 1940s.

A more complete explanation of the terminology and issues involved in these concepts appears in Chapter 3. Figure 1.11 shows that TPS was developed from 1945 and is still ongoing. It is a complex mix of changing attitudes, modifying procedures and equipment, while also taking into account the elimination of waste. The architect of the system, Taiichi Ohno, described waste as being one of seven basic types, each of which needed to be eliminated. The list appears as Figure 1.12 just for reference, with a more detailed explanation in Chapter 3.

It is interesting to compare and contrast the Six Sigma and TPS methodologies. Both systems have been implemented in many companies around the

Main Implementation Dates	Main Activity
1945–1955	Setup times reduced to 2–3 hours
1945–1974	Just-in-time introduced
1947–1949	Workplace redesign—parallel or L-shaped layout
1949–1950	Workplace redesign—horseshoe or rectangular layout
1949–1958	Intermediate warehouses abolished
1950–1955	Machine and assembly lines synchronized
1950–1955	Visual control system adopted on engine assembly
1953–1962	Supermarket system
1953–ongoing	Production leveling
1955–1961	Assembly and body plants linked
1962–1966	Full-work control of machines
1962–1971	Main plant setups reduced to 15 minutes
1962–ongoing	Kanban adopted company-wide
1963–ongoing	Workplace redesign—multiprocess operations
1966–ongoing	First automated line—Kamigo plant
1971–ongoing	Fixed-position stopping system in assembly

FIGURE 1.11
Implementation dates for some of the Toyota Production System (TPS) modules.

world, and each has offered many savings for the various companies which have implemented them. They approach quality from different perspectives.

The Six Sigma methodology states that its initial goal is 3.4 defects per million opportunities. Once a company is operating close to or at that level, the expectations are raised and the goal becomes Seven Sigma. An overriding tenet of the TPS system, however, is that the process is stopped

Overproduction
Time in hand (waiting)
Transportation
Processing
Inventory
Motion
Making defective product

FIGURE 1.12
Ohno's seven orders of waste.

as soon as a fault has been detected. In this scenario, no waste product or waste subassembly is generated. A solution to the problem is devised and implemented, and the process restarted.

Both approaches will work in driving out the "low-lying fruits" of waste reduction. In the long run, however, aiming for zero waste levels tackles more issues than aiming for minimum defects. Of course, the Motorola community had no intention of stopping waste reduction activities just because an arbitrary figure had been reached. Their long-term stated goal was always to keep going. Perhaps the introduction of their New Six Sigma program in the early part of the twenty-first century was Motorola's attempt to produce a more rounded quality system.

1.4 NEW SIX SIGMA

The New Six Sigma methodology is discussed in this whistle-stop tour of quality for the additional interesting issues that are now included. In their book *The New Six Sigma* (Barney and McCarty, 2003), the two principal architects suggest that their enhanced quality system contains the following elements:

- Six Sigma and shareholder value: key lessons from first-generation Six Sigma
- The four elements of New Six Sigma: Align, Mobilize, Accelerate, and Govern
- A step-by-step approach to New Six Sigma implementation and management
- Understanding the leadership governance enhancements that are key to success
- Refocusing Six Sigma tools on innovation and strategic business improvement
- Global case studies that demonstrate the effectiveness of the New Six Sigma
- Tomorrow's Six Sigma: new innovations in financial accounting, customer needs assessment, and asset management
- The future of the Six Sigma blackbelt

Some of these concepts are fascinating, plugging some of the soft skills gaps that were sadly missing in Six Sigma. On the other hand, they may not go far enough in their attempt to include current best practice from the business psychology viewpoint. Of equal interest is the discussion concerning the future of blackbelts, more of which later.

The four fundamental leadership principles of New Six Sigma—align, mobilize, accelerate, and govern—are described thus:

Align
- Using the performance excellence business model (based on the Malcolm Baldridge criteria), link customer requirements to business strategy and core business processes.
- Create strategy execution targets, and stretch goals and appropriate measures. The goal is to provide sustainable, measurable, bottom-line results that drive business goal achievement.

Mobilize
- Empower teams to drive improvements using projects selected by executives, project management methodology, and Six Sigma methodology.
- Organize team efforts with clear charters, success criteria, and rigorous reviews.
- Provide teams with just-in-time training and empower them to act.

Accelerate
- Employ an action-learning methodology by combining structured education with real-time project work and coaching to quickly bridge the gap from learning to doing. The motivation to act is perishable yet essential for driving projects to timely results.

Govern
- Drive the execution of strategy by managing scorecard metrics. Structured review processes involve reviewing dashboards of results as well as drilling into process and project details where needed. Barriers lift when leaders share best practices.

New Six Sigma recognizes many best practices, for example:

- Integrated business reviews
- Voice of the customer
- Business process redesign

- Scorecards
- Dashboards
- Blackbelt teams
- Blitz (*kaizen*) teams
- High-performance teams
- Focus on customers
- Cross-functionality

To my mind, New Six Sigma has many modules that are closer to TPS compared with the original Six Sigma. There is almost a coalescence of ideas, although the maturity and proven track record of TPS win the day. This initiative is nevertheless an attempt to update concepts and ideas from the initial quality system in order to make the quality system contemporary with today's business needs.

1.5 SO WHAT?

So we've had a look at a few quality systems, and we've left out a lot of material such as quality circles, TQM, TQC, and so on; what does this tell us that we didn't know already?

I think that the takeaway message is that the only constant in business is the need to change. What was good enough twenty years ago is not necessarily good enough today. Business models move on and so does the need for quality improvement, particularly in a shrinking global market. For example, as blackbelts take on an increasing commitment for quality training in other parts of their organizations, do they have the necessary skills in cultural differences? How does the green agenda that may well be enforced on companies, bailed out by central governments as part of the recent global financial restructuring, affect the concepts mentioned above? Can we just go on next year as we have continued to do this year? If we do, surely we will get what we got last time. So if we are all happy with our lot, post 2008 credit crunch, there is no reason to change. If, on the other hand, there is a growing customer need to focus on environmental responsibilities, plant footprint, cultural differences, and so on, then perhaps "the process" of process improvements needs to be reevaluated.

The other side of the coin is to do the best that we can do by implementing TPS in whatever form suits our companies, and to aspire to match the Japanese companies that have been developing and using these techniques for the last sixty years or so. I would suggest that the time may be ripe to evolve a system which has the potential to overtake their business models.

Now of course that is not going to be easy, and it will also take time. Additionally we have a quality system of training with all the checks and balances needed from the training accreditation perspective. Even more importantly, we have huge numbers of trained blackbelts who have gained their spurs at the coalface of continuous improvement. So the idea here is to suggest some enhancements on the continuing quality journey and not a disruptive technology that might force a change in direction. The basis of any change in approach must be evolutionary and not revolutionary, particularly in this case as there is a proven track record for TPS or lean thinking—whatever name the quality process uses in your respective organizations.

NOTES

1. http://www.juran.com.
2. http://www.qualityamerica.com/knowledgecente/articles/PYZDEKSixSigRev.htm.
3. http://www.motorola.com/content.jsp?globalObjectId=3074-5804.
4. http://www.motorola.com/content.jsp?globalObjectId=3079.
5. http://www.motorola.com/motorolauniversity.jsp.
6. http://www.sixsigmatraining.org/introduction-to-six-sigma/what-is-six-sigma.html

REFERENCES

Barney, M., and T. McCarty. 2003. *The New Six Sigma: A Leader's Guide to Achieving Rapid Business Improvement and Sustainable Results.* Upper Saddle River, NJ: Prentice Hall PTR, 1st edition. ISBN 0-131-01399-8.

Brue, G. 2002. *Six Sigma for Managers.* New York: McGraw-Hill, 1st edition. ISBN 0-071-38755-2.

Deming, W. E. 1986. *Out of the Crisis.* Cambridge: Cambridge University Press. ISBN 0-521-30553-5.

Harry, M., and R. Schroeder. 2006. *Six Sigma: The Breakthrough Management Strategy Revolutionizing the World's Top Corporations.* New York: Currency. ISBN 0-385-49438-6.

Pyzdek, T., ed. 2003. *The Six Sigma Handbook: The Complete Guide for Greenbelts, Blackbelts, and Managers at All Levels*. New York: McGraw-Hill, 2nd revised and expanded edition. ISBN 0-071-41015-5.

Whitmore, D. A. 1971. *Measurement and Control of Indirect Work* (The Heinemann accountancy and administration series). New York: Heinemann. ISBN 0-434-92252-8.

2

Production Systems

Before we start talking about production systems, it's worth a quick look at a definition. The *Encyclopaedia Britannica* defines them as follows:[1]

> All production systems, when viewed at the most abstract level, might be said to be 'transformation processes'—processes that transform resources into useful goods and services. The transformation process typically uses common resources such as labour, capital (for machinery and equipment, materials, etc.), and space (land, buildings, etc.) to effect a change. Economists call these resources the 'factors of production' and usually refer to them as labour, capital and land. Production managers refer to them as the 'five M's': men, machines, methods, materials and money.
>
> When viewed as a process, a production system may be further characterized by flows (channels of movement) in the process: both the physical flow of materials, work in the intermediate stages of manufacture (work in process) and finished goods; and the flow of information and the inevitable paperwork that carry and accompany the physical flow.

Basically, there are four common types of basic production systems:

- Batch system
  - General purpose equipment and methods are used to produce small quantities of output (goods or services) with specifications that vary greatly from one batch to the next. There are many examples of this type of system that include heavy-duty construction machinery, insurance claims, processed foods, specialized machine tools, and specialty chemicals.

- Continuous system
  - Items flow through a series of common operations, usually with a high throughput during which workers would undertake relatively small steps or segments. This type of system is often referred to as an *assembly line* that might be manufacturing automobiles, televisions, personal computers, or white goods. This type of system is noted for mass production of goods and/or services.
- Project system
  - This type of system is used for a "one-of-a-kind" product such as building a ship, a large computer, a bridge, or the like.
- Software production system
  - These are computer programs typically used to provide some form of artificial intelligence, which consists primarily of a set of rules about behavior. They might be known as *artificial intelligence systems* or *expert systems*. They need a database of rules and a rule interpreter that can respond to the relevant inputs.

In all four cases, there is a need to consider three basic requirements for the production system. These are

- Technology
  - Which is the most appropriate technology to use?
- Capacity
  - Will the system cope with current and future capacity should that go up or down?
- Adjustment mechanisms
  - How robust is the production system to external factors?

Production systems come in all shapes and sizes and are used to some extent or another in all facets of manufacturing and service organizations. Arguably, the first production system that is recognizable from today's perspective might be that of Newell and Simon's problem-solving technique of the 1950s,[2] the idea being written up in 1972 by Newell (Newell and Simon, 1972).

Historically, operations management was the catchall name used for a variety of business processes, including equipment maintenance, labor relations, management, principles of general management, production control, strategic policy, and production systems (Wild, 2002; Slack et al.,

1997). Arguably, the Toyota Production System now occupies some of the "management space" formally considered as "operations management."

The Toyota Production System (TPS) is a lot more than a simple computer program; indeed, many aspects of TPS do not require computers at all (Shingo, 1985, 1986). It is more of a sociotechnical system that comprises its management philosophy and practices. Through TPS, an organization can integrate supplier and customer interactions with the relevant logistics and manufacturing steps (Shingo, 1981, 1988). The underlying principles of TPS are documented in "The Toyota Way" thus:[3]

The Toyota Way is based on the Guiding Principles. Its five core values express the beliefs and values shared by the Toyota Group. All Toyota team members, at every level, are expected to use these values in their daily work and relations with others.

Challenge
    At Toyota, we maintain a long-term vision and strive to meet all challenges with the courage and creativity needed to realise that vision.
Kaizen
    Kaizen means striving for continuous improvement. As no process can ever be declared perfect, there is always room for improvement.
Genchi Genbutsu
    Genchi Genbutsu involves "going to the source to find the facts to make correct decisions, build consensus and achieve goals."
Respect
    Toyota respects others, makes every effort to understand others, accepts responsibility and does its best to build mutual trust.
Teamwork
    Toyota stimulates personal and professional growth, shares opportunities for development and maximises individual and team performance.

Arguably, the Toyota Production System is the basis for all of the modern continuous production systems. It is not always obvious, however, that TPS is in use in an organization because the production systems tend to be "badged" by the corporation or simply use the terms *lean, lean thinking*, or *lean Six Sigma*.

The need to create an in-house name and identity for what is essentially the Toyota Production System is interesting. In her book *Automotive Production Systems and Standardisation: From Ford to the Case of Mercedes-Benz* (Clarke, 2005), Constanze Clarke suggests that part of the problem, at least in the early days of TPS, was that without changing the name, there are constant reminders of the origins of the system:

> [T]he fact that the Chrysler Operating System had been modelled upon the TPS was a known fact within the automotive industry. Particularly for the German IG-Metal union, work councils and union representatives at Mercedes-Benz, the TPS was like a red rag to a bull. One key argument they raised was that the introduction of Toyota-based production system would result in a reduction of working cycles, job content, and an increase in repetitive work and physical and psychological strain: in short a revival of Taylorism. (p. 136)

This might be a generational issue. TPS concepts started to be implemented in the immediate post–World War II period. At that time, the concepts were introduced at the rate of one or two per year as the TPS developed. By the time that the full system was in operation in the 1970s, some of the original workers would have been replaced through natural wastage. At no time did the Toyota Motor Company have to implement all of the TPS concepts in a short space of time. The above comment from Clarke is interesting in that it highlights the concerns of the system that is now a way of life in Toyota.

There is also the issue of *not-invented-here*, or NIH. During my time in a multinational company, I experienced the same issue, although in those days NIH often related to ideas and potential products that did not originate in Munroe County (an area in the United States where my corporate headquarters was located).

Whatever the reasons for rebranding the TPS concepts, there are advantages to rolling out a production system across the company, some of which are as follows:

- A common language and training
- Common expectations
- A sense of shared understanding and purpose

These concepts are in addition to the more practical issues of workflow, waste management, inventory control, and the like. Standardized work allows reassigned workers and newcomers alike to more quickly contribute to the efficient running of the work cell or area.

Such has been the uptake of lean thinking concepts that there are implementations in a variety of industry sectors, including many unrelated to the motor industry. Eastman Kodak rebranded the concepts, although it kept much of the Japanese terminology, from *lean thinking* to the *Kodak Operating System* (KOS). A recent report from the Improvement and Innovation website, written by rathstrong_manager, posted the following article:[4]

The Kodak **Graphic Communications Group** (GCG) factory in Leeds, England is a model of lean in action. Bright yellow lines clearly de-lineate workspaces; tools are neatly arranged next to machines, their outlines painted so it is immediately visible if one is missing; everything is clean and tidy, there is no clutter, no waste materials; even the workers themselves are neatly turned out in matching safety gear. The whole place exudes calm efficiency.

The Lean Six Sigma journey for Kodak began in the 90s when the corporate head office introduced the **Kodak Operating System** (KOS). Based on the principles of Lean Six Sigma, it was implemented at the Leeds factory in 2002, with initially mixed results, as is often the case in the early days of an improvement program. However, Mike Harding, the Plant Manager at Leeds, believed that it was the right thing for the site. 'We knew what we wanted to do but didn't know how to do it,' he says. 'These methodologies gave us the tools to do it.'

The focus of the program changed from having a handful of projects run by two or three blackbelts, to a more strategic approach employing a large number of greenbelts working on many projects, with training and support from consultants Rath & Strong. This proved much more successful, lifted the organisation and the program began to take off.

This report and the quotes from Mike Harding are interesting in that they suggest that they were successful in employing a large number of greenbelts rather than a few blackbelts. The issue of training and "belt" terminology will be covered in Chapter 6. I have known Mike Harding for over twenty years and know that there were many months' (perhaps even years') worth of hard work before the project was deemed a success. Nevertheless, the patience and hard work paid off.

During his tenure as chairman and CEO of Eastman Kodak, George Fischer introduced the requirement to align activities to the benefit of customers, employees, or shareholders. This alignment was embedded in the Kodak Management Systems as follows:

- Customer satisfaction
- Employee satisfaction
- Shareholder satisfaction

Incidentally, the Kodak Operating System also has KODAK Values associated with the culture change. They can be found on the Kodak website[5] and below:

- *Respect for the dignity of the individual.* We cannot operate effectively unless each of us is able to treat everyone else with appropriate respect. This essential value is at the heart of our culture and will help us focus on many important issues like diversity of our workforce.
- *Integrity.* In today's increasingly complex business and social world, integrity and honesty must be the hallmarks of any organization or person striving to consistently achieve and maintain the respect of our publics.
- *Trust.* We must be able to work in an environment in which we trust each other. We must depend upon and trust our colleagues to do their assigned tasks without the need to check and recheck their work. Likewise, each of us must handle our responsibilities so that our colleagues can trust we are doing our part.
- *Credibility.* Each of us must earn credibility with others inside and outside the company. Certainly, the company as a whole must strive for the highest credibility with all its external publics. We must commit to do what we say we will do, and no later than we commit to do it.
- *Continuous improvement and personal renewal.* Results do count, and continuous improvement toward world-class levels is essential to achieve credibility with our publics. We must each continually improve ourselves and renew our skill sets. Training and education must be accepted as a common responsibility between us as employees and the company as a whole.
- *Recognition and celebration.* We will search out and welcome opportunities to openly celebrate the achievements of others and congratulate individuals, teams, employees, suppliers, and customers for

delivering results that contribute to Kodak success. Recognition and celebration will be integral parts of our everyday work activity.

In many ways, these values share common ground with the Toyota Way. Unfortunately, during my time at Kodak we were unable to fully integrate all of the TPS concepts so that they became a way of life for the whole site of their primary UK manufacturing operation. More through technology innovation than anything else, this site may well close in the next year or two. This closure will be the direct result of shifting markets from traditional to digital technology and nothing to do with the partial implementation of KOS. *All* of the people implementing KOS worked tirelessly and to very high standards.

This is just one of literally tens of thousands of successful lean thinking implementations across many industrial sectors. The finance sector has been equally vigorous in their implementation of the concepts. Unfortunately in these post-2008 credit crunch days, which were caused by a small community within the finance sector ignoring the effects of so-called toxic debt, implementations in the finance sector have been overshadowed by other issues.

Given that lean thinking started in the Toyota Motor Company, it might be instructive to take a quick look at some of the other motor manufacturers. General Motors offers the following comments on its Canadian website:[6]

The **GM Global Manufacturing System (GMS)** is an important building block of an integrated strategy to develop products that excite our customers in markets around the world. GM is bringing together the best, most competitive manufacturing practices from around the world and leveraging what it has learned as it moves to a common global manufacturing system for all of its new plants and existing facilities. The system is dynamic. With each new plant or renovation of a current plant, it is further refined and implemented with regional variation, based on the individual plant environment, supplier capability, vehicle architecture and cultural factors.

At the heart of the system is the operator in the plant - the person who builds GM's great products. Plants and processes are designed around providing support for the operators and teams on the plant floor, so they can efficiently build great vehicles that provide our customers with higher quality, value and responsiveness.

Several factors played a role in the evolution of GM's GMS. The experience gained through NUMMI, a joint venture with Toyota provided the introduction of the Toyota Production System techniques into GM. Continuous focus is placed on research of best practices of GM's competitors and joint venture partners, along with forging closer ties with suppliers to include a much greater willingness to accept alternate or new ideas.

Manufacturing performance is improved through the consistent adoption of five principals—people involvement, standardization, built in quality, short lead time and continuous improvement. The principals are interrelated and implemented as a complete system.

This quality initiative clearly helped to make General Motors a more successful organization, at least from the quality perspective, for it has recently won awards for its product quality from J.D. Power and Associates.[7] Unfortunately, General Motors was also a high-profile casualty of the 2008 global financial crisis; only time will tell if they will continue to trade as GM in the longer term.[8]

Many if not all of the motor manufacturers have embraced lean thinking concepts to one degree or another. Nissan, however, also mentions its six objectives of "Nissan GT 2012: Quality Leadership." They are[9]

- to halve warranty claim rates for the first three months in service after delivery.
- to halve the supplier parts defect rate.
- to halve the breakdown ratio compared with FY2007.
- to halve the lead time from occurrence of defect to preparation of new parts.
- to double the regions where SSI (Sales Satisfaction Index) and CSI (Customer Service Index) are on the top level.
- to double the number of models which are rated high in perceived quality by both internal and external metrics.

So perhaps not all of the lean thinking concepts have been implemented in their entirety; otherwise, Nissan might not have needed to implement its GT 2012 program.

Let's take a look at a few of the global top ten motor manufacturers so that we can try to assess the impact of implementing production systems.

2.1 TOP TEN MOTOR MANUFACTURERS AND VEHICLE PRODUCTION

There has to be a common currency when assessing global companies, and it tends to be the U.S. dollar. Although figures are available on some of the top manufacturers' websites for more recent company accounts, not all of the data are available for 2008.[10] It is not such a bad thing to look at the global motor manufacturers before this current credit crunch hit us as we can have a look at how the companies were performing without external pressures from these abnormal trading and financial constraints.

There are always plenty of caveats when looking at this sort of information since not all of these companies sell exactly the same types of vehicles. The Toyota Motor Company, for example, sells almost 1 million hybrid or dual-fuel cars. Clearly, the largest motor manufacturer by revenues *and* by profits is the Toyota Motor Company, which was only established in the 1930s.

These motor manufacturers are supported by suppliers for all of the components (see Figure 2.2).

If you add together the revenues for the top ten manufacturers and the top ten suppliers, the number was over $1,600,000 million for 2007 (where $1 million is $1,000,000). Even Texan oil barons will think this number

2007 Company	Revenues $ Million	Profits $ Million	% Change from 2006	Assets $ Million
Toyota Motor	230,201	15,043	7	326,099
General Motors	182,347	−38,732	0	148,883
Daimler/Chrysler	177,167	5,446	35	197,516
Ford Motor	172,468	−2,723	0	279,264
Volkswagen	149,054	5,639	64	212,521
Honda Motor	105,102	5,254	4	126,745
Nissan Motor	94,782	4,223	7	119,952
Peugeot	82,965	1,211	449	100,846
Fiat	80,112	2,673	100	87,923
BMW	76,675	4,279	19	130,119

FIGURE 2.1
The top ten global motor manufacturers by revenue (2007).

Company	Revenue $ Millions	% Change from 2006
Robert Bosch	63,401	16
Denso	35,245	14
Johnson Controls	34,678	7
Bridgestone	28,793	12
Delphi	26,160	−1
Aisin Seiki	23,646	16
Michelin	23,087	12
Continental	22,748	22
Goodyear Tire & Rubber	20,538	1
ZF Friedrichshafen	17,314	18

FIGURE 2.2

The top ten motor industry suppliers by revenue.

large! Of course, this extremely large number is also part of the problem. For no matter how financially viable these individual corporations are, governments *have* to help bail them out in times of world recessions because this industry represents countless thousands of jobs across the world. These companies also pay hefty taxes, providing a further incentive for governments.

Several observations can be made from the global motor manufacturers shown in Figure 2.1. The first is the scale of the problems at General Motors and Ford Motor Company[11] (see Figure 2.3). The high-level net profit and loss as expressed in the annual reports to shareholders, found on their respective websites, suggest that neither GM nor Ford have enjoyed healthy profits for a number of years, and both sought large government handouts in the first quarter of 2009!

This book is not about berating the various senior managers in the automotive industry. The actual purpose is to highlight that their systems for

Company	Net Profit and Loss				
	2007	2006	2005	2004	2003
GM	−38,732	−1,978	−10,417	+2,804	+3,859
Ford Motor Company	−2,723	−12,600	+2,000	+3,500	+500

FIGURE 2.3

Net profit and loss for GM and Ford, 2007 and 2003.

Toyota Motor Company	Net Profit and Loss				
	2008	2007	2006	2005	2004
Net revenues (¥ billion)	26,289	23,948	21,036	18,551	17,294
Operating income (¥ billion)	2,270	2,238	1,878	1,672	1,666
Net income (¥ billion)	1,717	1,644	1,372	1,171	1,162
Return on equity	14.5%	14.7%	14.0%	13.6%	15.2%

FIGURE 2.4
Net profit and loss for the Toyota Motor Company.

dealing with high waste and lower sales or a combination of the two have scope for improvement. As stated above, if that wasn't happening in the good times, how is it going to happen in a global recession?

By contrast, the equivalent data for the Toyota Motor Company[12] are shown in Figure 2.4. Return on equity (ROE) is often defined as the ratio of net income to shareholder's equity. Using this definition, net income is for the full fiscal year before dividends are paid to common stockholders, but after dividends are paid to preferred stock. Shareholders' equity does not include preferred shares. This ratio basically shows how much profit a company had made with the money shareholders have invested. Interest rates on investments, at least in the United Kingdom, rarely exceeded 10 percent. A figure for ROE greater than 14 percent represents a very sound investment. By the way, during this period, Toyota vehicle production rose steadily (Figure 2.5).

In all years, Toyota sold all of the vehicles that it manufactured. In other words, its remaining inventory stock was zero—Toyota was making to order. The profit and loss figures discussed in this section don't tell the whole story for the global industry as there were successes. Let's have a look at a few press releases from some of the aforementioned companies, posted before the 2008 subprime financial crisis:

	2008	2007	2006	2005	2004
Japan	5,160	5,100	4,684	4,534	4,284
Overseas	3,387	3,080	3,027	2,697	2,229
Total	8,547	8,180	7,711	7,231	6,513

FIGURE 2.5
Toyota vehicle production in thousands of units.

Nissan announced the addition of 800 staff and implemented a third production shift at its manufacturing plant in Sunderland to meet demand for the QASHQAI (a model of car). Since its European launch in March 2007, sales have reached 130,000 units and, as a result, boosted production at Sunderland by 20%.

Honda has announced £80 million investment in its Swindon plant, upgrading the paint shop and plastics operation, taking total investment in Swindon to £1.38 billion. It includes £16 million for casting diesel engine blocks, a process only Honda Japan has undertaken to date. The announcement came as the two-millionth Honda left the Swindon production lines. Operations began 22 years ago and today the plant produces nearly 240,000 cars a year.

Toyota will produce a new petrol engine at its plant in Deeside, North Wales. This represents an additional investment of £88 million and production of the new engine is scheduled for late 2009. In February 2007 the plant celebrated the production of its three millionth engine. This is the second major investment in engine production at the plant since 2007, and is in addition to the £700 million already invested in Toyota's operation.

Similar successes were posted by Leyland Trucks, Vauxhall, Mini, Ford, and Rolls Royce. In global terms, the UK market is fairly small as it represents about 3.1 to 3.7 percent of the global passenger car production. Nevertheless, the value of these exports accounted for an average over the previous five years of around £22 billion (where 1 billion = 1,000 million).

While there were markets buoyant in 2007, the picture changed in 2008, no more so than in the U.S. market (Figure 2.6), where the data represent millions of units sold. The shifts in the numbers are dramatic and don't add up! The reason why they don't add up is that there are millions of imported units reaching the United States every year. That in itself highlights another issue concerning the quality of goods produced and sold. Unless the U.S. manufacturers can produce vehicles to the same levels of quality and cost as the imported vehicles, there is little to stop consumers

	2008	2007
Total U.S. production	8.7m	10.8m
Total U.S. sales	13.2m	16.2m

FIGURE 2.6
Total U.S. production and sales of vehicles for 2007 and 2008.

	February 2008 (in Numbers of Units)	Ratio of Domestic to Imports 2008	February 2009 (in Numbers of Units)	Ratio of Domestic to Imports 2009
Domestic cars	391,080		219,907	
Import cars	168,200	2.325	120,986	1.818
Total cars	559,280		340,893	
Domestic light trucks	511,187		271,116	
Import light trucks	101,214	5.051	75,347	3.598
Total light trucks	612,401		346,463	
Domestic light vehicles	902,267		491,023	
Import light vehicles	269,414	3.349	196,333	2.501
Total light vehicles	1,171,681		687,356	

FIGURE 2.7
U.S. domestic sales versus imports.

from moving to more reliable brands *or* to vehicles that better suit the mood of the buyer.

For example, and without the aid of a crystal ball: the global recession will force some green issues such as hybrid cars and exhaust emissions. The Toyota Production System allows for a switch to smaller engine sizes within the process of car manufacture. So smaller engines can be ordered whenever needed, and the car built to expectations. Arguably, the Toyota Motor Company also has a more established track record for hybrid cars. So if we look at the U.S. sales of domestic cars, light trucks, and light vehicles (Figure 2.7), we can see that the ratio of domestic to import was higher in 2008 compared with 2009, at least for the snapshot comparison from February 2008 to February 2009.

So what other threats are out there for the global markets? Well, interestingly, Tata Motors of India[13] has generated a huge frenzy for its new model, the Nano, which retailed at £1,400 in March 2009. There has been so much interest that lotteries now take place with the lucky winners able to buy one of the cars. Now, these vehicles are not high-spec cars (they are built as three models: basic, air conditioned, and air conditioned with powered front windows), and Tata will struggle to make sufficient units to service their own domestic market. However, they do act as disrupters for the rest of the global markets in several ways, not least because other manufacturers will also come out with low-cost vehicles.

Additionally, these units will affect the secondhand car market in India (such as it is).

The potential market erosion of 0.5 to 1 percent over time might well become a few percentage points within a few years. It could be similar to the IBM–Microsoft issue concerning the DOS operating system when IBM led the computer world but nevertheless allowed Microsoft into the software market.

In this economic climate, *any* market share reduction should be taken seriously.

The other issue of note is that of overcapacity within the global markets—in other words, the number of vehicles produced as a function of the number of potential customers. With the exception of the Tata Nano and indeed the high-spec luxury end of the market, there is a general overcapacity in the marketplace and has been for a number of years. There are various factors affecting customer choice, one being price and another reliability; both are affected by the quality policy, procedures, and practice.

NOTES

1. http://www.britannica.com/EBchecked/topic/478032/production-system.
2. http://www-formal.stanford.edu/jmc/generality/node3.html.
3. http://www.toyota-forklifts.co.uk/En/company/Pages/The%20Toyota%20Way.aspx.
4. http://www.improvementandinnovation.com/features/project/kodak-operating-system-successfully-integrating-lean-and-six-sigma.
5. http://www.kodak.com.
6. http://www.gmcanada.com/inm/gmcanada/english/about/Quality/quality_global.html.
7. http://businesscenter.jdpower.com.
8. http://www.gm.com.
9. http://www.nissan-global.com/EN/NEWS/2008/_STORY/080924-02-e.html.
10. http://money.cnn.com/magazines/fortune/fortune500/2009/.
11. http://corporate.ford.com/investors/reports-financial-information.
12. http://www.toyota.co.jp/en/ir/financial/high-light.html.
13. http://www.tatamotors.com.

REFERENCES

Clarke, C. 2005. *Automotive Production Systems and Standardisation: From Ford to the Case of Mercedes-Benz (Contributions to Management Science)*. Heidelberg: Physica-Verlag. ISBN-10: 3-790-81578-0, ISBN-13: 978-3790815788.

Newell, A., and H. Simon. 1972. *Human Problem Solving*. Englewood Cliffs, NJ: Prentice Hall. ISBN-10: 0-134-45403-0, ISBN-13: 978-0134454030.

Shingo, S. 1981. *Study of "Toyota" Production System from the Industrial Engineering Viewpoint*. Tokyo: Japanese Management Association. ASIN B0007BSQQW.

Shingo, S. 1985. *A Revolution in Manufacturing: The SMED System*, trans. A. Dillon. New York: Productivity Press. ISBN-10: 0-915-29903-8.

Shingo, S. 1986. *Zero Quality Control: Source Inspection and the Poka-Yoke System*. New York: Productivity Press. ISBN-10: 0-915-29907-0.

Shingo, S. 1988. *Non-Stock Production: The Shingo System for Continuous Improvement (Most Detailed Examination of the Fundamentals of Jit)*. New York: Productivity Press. ISBN-10: 0-915-29930-5.

Slack, N., S. Chambers, A. Harrison, and C. Harland. 1997. *Operations Management*. London: Financial Times/Prentice Hall. ISBN-10: 0-273-62688-4.

Wild, R. 2002. *Essentials of Operations Management*. Andover, UK: Thomson Learning. ISBN-10: 0-826-45271-X.

3

The History and Development of the Toyota Production System

It is worth spending a few minutes looking at the origins of the Toyoda Spinning and Weaving Company since it was from the profits of this company that the Toyota Motor Company was founded. Some of the inventions from the weaving industry were, at the time, revolutionary. Indeed, as recently as 1985, the Japanese Patent Office listed Sakichi Toyoda (Toyoda Spinning and Weaving Company's founder) as one of the ten most important inventors in Japanese history. Figure 3.1 lists "Sakichi's Patents" for further reference as they contain some interesting concepts. Figure 3.2 is a timeline of significant events in Sakichi's life.

Throughout his youth, Sakichi gave his son Kiichiro Toyoda the opportunity to develop his own flair for business and technology, so much so that just prior to his death, Sakichi told Kiichiro "to have your own life's work" and that he believed in the automobile concepts that were being developed by Kiichiro. This was despite the fact that General Motors and Ford dominated the market in Japan at that time (with 84 percent of the Japanese market).

The death of Sakichi in 1930 led Kiichiro to invest his time, his energy, and some of the profits from the weaving business into the development of the Toyota Motor Company. The fledgling motor company further benefited from the weaving company legacy in that Taiichi Ohno, who went on to develop the Toyota Production System, was himself an employee of the Toyoda Spinning and Weaving Company prior to joining the Toyota Motor Company in 1943.

Title	Patent Number	Date
Loom mechanism	CA282836 (A)	1928-08-28
Loom mechanism	CA282835 (A)	1928-08-28
Shedding motion for circular looms	US1691314 (A)	1928-11-13
Taking-up device for circular looms	US1689507 (A)	1928-10-30
Dispositif d'enroulement pour métier à tisser circulaire	FR646744 (A)	1928-11-15
Mouvement de formation du pas pour métiers à tisser circulaires	FR646743 (A)	1928-11-15
Fachbildevorrichtung für Rundwebstühle	CH129854 (A)	1929-01-02
Abzugvorrichtung für Rundwebstühle	CH129257 (A)	1928-12-01
Mechanism for automatically changing shuttles in looms	GB191013908 (A)	1911-06-08
Improvements in heddles for looms	GB190817684 (A)	1909-01-28
Improvements in or connected with loom-picking mechanism	GB190912160 (A)	1909-08-26
Automatic shuttle-changing mechanism for looms	GB190811386 (A)	1909-05-26
Tension and let-off for looms	GB190807085 (A)	1908-07-16
Improvements in circular looms for weaving	GB190702845 (A)	1907-10-31
Webstuhl mit kreisförmig angeordneter Kette	CH40020 (A)	1908-06-01
Fachbildungsvorrichtung für Rundwebstühle	AT37752 (B)	1909-06-25
Väv til rörformet Vare	DK10902 (C)	1908-06-09

FIGURE 3.1
Sakichi's patents.

3.1 THE TOYODA SPINNING AND WEAVING COMPANY

Sakichi Toyoda formed the Toyoda Spinning and Weaving Company in 1918. His first invention allowed a productivity increase of 50 percent compared with other indigenous looms in use at that time. This first Toyoda power loom found initial sales success with small manufacturers weaving narrow cloth for such export markets as Korea, Manchuria, and Taiwan. Arguably, his most important single achievement was a patented let-off device that maintained the warp at a constant tension as it was being fed off the warp beam.

Sakichi was a born innovator as well as an inventor; he was always looking for new ideas or hired people who could deliver new ideas. He hired Charles A. Francis, an American teacher of mechanical engineering who was at that time at the Tokyo Higher Technical School, and

1867	Born into a family of carpenters and grew up in Shizouka, Japan.
1887	Began work on looms.
1894	Birth of Kiichiro (Sakichi's son, who founded the Toyota Motor Company).
1895	Founded Toyoda Co. to sell yarn-reeling machine.
1899	Mitsui & Co. signed a ten-year agreement with Sakichi for rights to his power loom.
1907	Dissolved Toyoda Co. and created Toyoda Loom Works.
1910	Resigned from Toyoda Loom Works (but the company kept the name).
19??	Went back to work on looms; raised his own financing. Created Toyoda Automatic Weaving.
1918	Toyoda Automatic Weaving became Toyoda Spinning and Weaving Company.
1926	Sakichi founded Toyoda Automatic Loom Works.
1929	Toyoda Automatic Loom Works, Toyoda Spinning & Weaving, Toyoda Spinning & Weaving Works in Shanghai.
1930	Agreement between Toyoda and Platt to make the latter's looms.
1930	Died on 30 October from pneumonia.

FIGURE 3.2
Sakichi Toyoda: A timeline.

who had also been employed as an engineer for the Pratt and Whitney Company. From 1905 to 1907, Francis trained workers in the basic techniques of machine manufacture including the use of indicators and gauges that he had been exposed to during his previous employment history. Francis also introduced the company to batch production of standard models. He taught engineers about the design of jigs and fixtures and the layout of equipment on the production line, and he advised managers on essential high-quality machine tools to consider purchasing.

A profound feature of Sakichi's loom was that it stopped instantly if any of the warp or weft threads broke, producing no waste in the process. Essentially, this loom had the capability to distinguish between normal and abnormal operating conditions. In 1930, Sakichi sent Kiichiro to England to broker a deal with Platt Brothers, who purchased the patent rights for the loom for 1 million yen—a small fortune in those days. In itself, this was just another business deal; in practice, it proved farther reaching as the funds were spent on automobile research and development. Kiichiro announced the goal to develop cars produced in Japan for the general public in 1933 thus:

[W]e shall learn production techniques from the American method of mass production. But we will not copy it as it is. We shall use our own research and creativity to develop a production method that suites [*sic*] our own country's situation. (Ohno, 1988)

Four years later, Kiichiro founded the Toyota Motor Company; the name *Toyota* was created from Toyoda by the automobile division for marketing purposes.

After graduating from the Department of Mechanical Engineering of Nagoya Technical High School in 1937, Taiichi Ohno joined Toyoda Spinning and Weaving. In his book *Toyota Production System: Beyond Large-Scale Production* (Ohno, 1988), he makes the following observation:

[I]n 1937 I was working in the weaving plant at Toyoda Spinning and Weaving. Once I heard a man say that a German worker could produce three times as much as a Japanese worker. The ratio between German and American workers was 1 to 3. This made the ratio between Japanese and American workforces 1 to 9. I still remember my surprise at hearing that it took nine Japanese to do the job of one American. (p. 3)

By 1937, the scene was set. There was an industrialist with a deep interest in automobiles with a history of manufacturing process innovation. Additionally, in Taiichi Ohno there was an extremely able worker who would go on to influence many aspects of the fledgling motor company with manufacturing concepts that are now changing aspects of manufacturing throughout the industrialized world.

3.2 THE TOYOTA MOTOR COMPANY

Taiichi Ohno took the concept of producing zero defects with him on his transfer to the Toyota Motor Company in 1943 and developed the waste reduction concepts one by one. He enlisted the help of Shigeo Shingo to realize the concepts that became known as *poka-yoke*, or mistake proofing, and *single-minute exchange of die*, which will be explained in this chapter. Ohno was named machine shop manager in 1949, senior managing director in 1970, and vice president in 1975. Over his thirty years or so at the Toyota Motor Company, Ohno was to revolutionize the motor manufacturing in Toyota. Ohno's influence has since spread to many other corporations in the automotive industry, including suppliers, and latterly to many other business sectors. Figure 3.3 shows a timeline of the concepts that were introduced as part of the Toyota Production System credited to Taiichi Ohno; those credited to Shigeo Shingo will be covered later.

The timeline on the following page shows that the system not only is complex but also took thirty years to develop to the point where most aspects of the current system are recognizable. Up until the early 1970s, few people outside of Toyota knew of the work that had taken place. The ability to react to customer needs, however, brought the company to worldwide prominence from 1973 onward. Those of us old enough to remember the oil crisis of 1973 will remember the high prices demanded by the Middle Eastern oil producers. A manufacturing system that allowed customers to order cars of smaller engine sizes ensured that Toyota made only the cars that the customers wanted. Of course, they had no stocks of completed vehicles that were unwanted, and so their recovery from the crisis was swift, unlike those other car manufacturers which had vast stocks of unwanted vehicles.

The success of the production system during this crisis led to its implementation company-wide, with Ohno promoted to executive vice president in 1975. He retired from regular service with the company in 1978 and returned to his roots in the weaving company for his remaining few working years. As we have seen, since that time the Toyota Motor Company has gone from strength to strength and is now the largest motor manufacturer in the world.

While Figure 3.3 shows a timeline of the introduction of various components of the Toyota Production System (TPS), what is it and how does it work? Well, there are whole books and many courses designed to instruct the novice on the finer aspects of the system. Perhaps a few of the more fundamental and/or interesting concepts should be briefly explained here just to show the breadth and depth involved in TPS.

The use of continuous incremental improvements, known as *kaizen*, was adopted in the deployment of the methodologies discussed here. Three concepts were introduced during the implementations discussed in this chapter, and they deserve definition here:

Year	Technique Introduced
1945–1975	Just in time
1945–1955	Setups (2–3 hours)
1947	Two-machine handling (parallel or in L-shaped layout)
1948	Withdrawal by subsequent processes ("upstream" transport)
1949–1950	Three- or four-machine handling (either horseshoe or rectangular layout)
1949	Intermediate warehouses abolished
1950	Machining and assembly lines synchronized
1950	Visual control, *andon* system adopted in engine assembly
1953	Call system for the machine shop
1953	Production leveling introduced
1953	Supermarket system in machine shop
1955	Assembly and body plants linked
1955	Main plant assembly line production system (*andon*, line stop, and mixed load)
1955	Required number system adopted for supplied parts
1955	Whirligig water system (small load and mixed transportation)
1957	Procedural chart (*andon*) adopted
1958	Warehouse withdrawal slips abolished
1959	Transfer system (in/in and out/out)
1961	*Andon* installed, Motomachi assembly plant
1961	Pallet *kanban*
1961	Red and blue card system for ordering outside parts
1962	Full-work control of machines, machine *baka-yoke*
1962	*Kanban* adopted company-wide
1962	Main plant setups (15 minutes)
1963	Multiprocess operation
1963	Use of interwriter; system of autonomation selection of parts adopted; information indicator system adopted
1965	*Kanban* adopted for ordering outside parts, 100% supply system
1965	Began teaching Toyota system to affiliates
1966	First automated line, Kamigō plant
1971	Body indication system (Motomachi Crown line)
1971	Fixed position stopping system in assembly
1971	Main office and Motomachi setups (3 minutes)
1973	Transfer system (out/in)

FIGURE 3.3

Timeline of Toyota Production System (TPS) techniques.

- *Heijunka*, or production leveling, is the concept that leads to goods being produced for the customer when the customer wants them. By way of an example, let us assume that there is a market need for one thousand units of product in each of four different colors once per month. If there are twenty-five days in the working month, then forty units will be made in each day, ten of each color. In the Toyota Production System, one car will be made to each color—for example, red, green, yellow, then blue—assuming that the orders come into the factory in sequence. The cycle will be repeated until forty units are manufactured in that working day. At that point, production ceases even if there is spare capacity. If there is spare capacity, the workforce will be redeployed to do other job functions. If there continues to be a workforce surplus, there will be some workforce redeployments until the correct staffing levels are obtained. In order to avoid excessive daily workforce redeployment, production leveling is taken as far upstream in the manufacturing process as possible.
- *Takt* time is defined as the length of time in seconds or minutes that is taken to produce one piece of product; it is calculated by taking the length of time a machine is producing product and dividing it by the number of units produced during that period.
- *Jidoka* is the concept of automatic line stopping whenever a defect is detected so that no bad parts will be passed forward to interrupt the downstream flow of materials. As far as possible, this process is automated. In the extreme case there is no quality assurance department or final product-testing department, as there should be no defective products.

The management team at the Toyota Motor Company has driven this process for the last few decades and worked with its suppliers for years helping them to introduce the same principles. The complete analysis of waste using TPS makes the following two points:

- Improving efficiency makes sense only when it is tied to cost reduction. To achieve this, we have to start producing only the things we need using minimum manpower.
- Look at the efficiency of each operator and of each line. Then look at the operators as a group, and then at the efficiency of the entire plant

(all the lines). Efficiency must be improved at each step and at the same time for the plant as a whole.

These guiding principles led Ohno to describe waste as being one of the following seven types, as mentioned in Chapter 1:

1. Overproduction
2. Time in hand (waiting)
3. Transportation
4. Processing
5. Inventory
6. Motion
7. Making defective product

The Japanese word for waste is *muda,* and so this list is sometimes referred to as *Ohno's seven orders of muda.*

Time in hand or waiting is sometimes difficult to detect or indeed prevent. We are all sometimes guilty of filling in our time with "jobs." The regulation of time in this context (waiting) led to the introduction of a *kanban,* which is used to regulate workflow. The rules for a *kanban* appear in Ohno's book (1988) and as Figure 3.4.

Function of a *Kanban*	Rules for Use
Provides pick-up or transport information	Later process picks up the number of items indicated by the *kanban* at the earlier process.
Provides production information	Earlier process produces items in the quantity and sequence indicated by the *kanban.*
Prevents overproduction and excessive transport	No items are made or transported without a *kanban.*
Serves as a work order attached to goods	Always attach a *kanban* to the goods.
Prevents defective product by identifying the process making defectives	Defect products are not sent on to the subsequent process. The result is 100% defect-free goods.
Reveals existing problems and maintains inventory control	Reducing the number of *kanbans* in a production line increases their sensitivity.

FIGURE 3.4
Kanban rules.

Preventing defective products from being sent to a subsequent process is in sharp contrast to the Six Sigma concepts that were being developed by Motorola in approximately the same time frame. The adaptation of the concepts developed by Toyoda Sakichi while developing his auto-activated loom to the production of cars is much in evidence here.

Another concept that is thrown into doubt by *kanbans* is that of large dedicated facilities used in mass production (see, e.g., Wild, 1989). Of increasing importance is the use of smaller yet infinitely more versatile production units that can handle small quantities of materials. This leads on naturally to continual workplace redesign, and the need to develop a different approach to planning both machine utilization and the workforce.

The driving force for these changes is workforce and cost reduction. Ohno made the following comment:

> In the Toyota production system, we think of economy in terms of manpower reduction and cost reduction. The relationship between these two elements is clearer if we consider a manpower reduction policy as a means of realizing cost reduction, the most critical condition for a business's survival and growth. Manpower reduction at Toyota is a company-wide activity whose purpose is cost reduction. Therefore, all considerations and improvement ideas, when boiled down, must be tied to cost reduction. Saying this in reverse, the criterion of all decisions is whether cost reduction can be achieved. (1988, p. 53)

Ohno returned to the theme of labor reductions later on in his book, stating that

> 'using fewer workers' can mean using five or even three workers depending on the production quantity[;] there is no fixed number. 'Labour saving' suggests that a manager hires a lot of workers to start with, reducing the number when they are not needed. 'Using fewer workers', by contrast[,] can also mean working with fewer workers from the start. (1988, pp. 67–68)

Ohno was only too aware of an issue involving labor numbers, as there had been a bitter dispute between the union within Toyota and the senior management team. The Japanese economy was suffering from runaway inflation in 1949, and one of the countermeasures was to reduce available commercial credit until credit dried up completely. Toyota's financial situation deteriorated to such an extent that the company was unable to pay its payroll. Senior management decided to reduce workforce numbers, leading to a bitter strike which started in April 1949. A solution was finally reached between management and the union, and workforce numbers reduced from 8,000 to 6,000 employees. President Kiichiro Toyoda and his executive staff resigned at this time.

An important element of the Toyota Production System has not yet been discussed, that of *single-minute exchange of die* (SMED). It has been separated from the earlier discussions as it was developed by Shigeo Shingo—a consultant who worked at various Toyota facilities as well as for some subsidiary companies and some suppliers. In an article called "Bringing Wisdom to the Factory" in Shingo's book *A Study of the Toyota Production System* (1989), Taiichi Ohno explained,

> This was uneconomical and therefore unacceptable. Since we also wanted maintenance to be done during working hours, we began to study the question of how setup changes could be performed in a very short period of time. Shigeo Shingo, of the Japan Management Association, was advocating 'single-minute setup changes' and we felt that this concept could be of great service to us. It used to be that after spending half a day on setup, the machine might be used for only ten minutes. Now one might think that since setup took half a day, production ought to continue for at least that long. This would have left us with a lot of finished products we could never sell.

It took nineteen years to develop SMED. A brief timeline appears as Figure 3.5.

Shingo realized that in setting up any machine, there are two fundamentally different types of setups:

Year	Plant
1950	An improvement study for Toyo Kogyo's Mazda plant in Hiroshima
1957	Improvement to a diesel engine bed planer at Mitsubishi Heavy Industries
1969	Setup time reduction for a 1,000 ton press at Toyota Motor Company

FIGURE 3.5
Single-minute exchange of die (SMED) timeline.

- *Internal setup* (IED), such as mounting or removing dies, that can only be performed once the machine has stopped
- *External setups* (OED), such as transporting equipment, that could be conducted while the machine is in operation

Prior to an exposure to these concepts, most manufacturing companies used a mixture of IED and OED to such an extent that it was difficult for the local management and operations personnel to separate out the tasks in their own minds. Moving operations from IED to OED might require investment or a total rethink of the operations and engineering tasks associated with a setup, but is well worth the effort (Liker, 1997).

The SMED techniques proved such a success at Toyota that they became part of the Toyota Production System. In his book *A Revolution in Manufacturing: The SMED System* (Shingo, 1985), Shingo outlines many examples from different companies around the world that have benefited from SMED techniques. Such examples show the versatility of this technique, which has now been applied to situations unrelated to manufacturing, including Formula 1 racing.

The operational consequences of the implementation of SMED often lead to a decrease in skill level needed to perform what may have once been a complicated product changeover. If this is acceptable to the workforce, then SMED techniques will be embraced. If the engineers feel as if their employment prospects will become more limited, then there will be an adverse effect. One of the senior managers in the Citroën plant in France made the following observations when SMED was introduced:

Lately, though, machine operators have been able to perform setup changes themselves and they spend a month at a time concentrating on the operation of a single machine. This gives them a feeling of responsibility for their machines. I'm sure that's why they're now oiling and taking care of their equipment. (Shingo, 1985, p. 118)

Xavier Carcher, vice president of Citroën, continued in the same vein,

[S]ince the success of SMED, though, there's a new determination to come up with ways to make them work; the emphasis is on putting ideas into practice. What's more, when someone who comes up with a suggestion is told why his idea won't work, he changes his tack and makes suggestions.

This has speeded up conferences, and a number of problems have been resolved by putting suggestions into practice. With so many improvements, productivity has recently risen substantially. The thing I have been most grateful for has been the revolution in everyone's attitudes towards improvement. (Shingo, 1985, p. 124)

Shingo is also credited with developing the concept of *zero quality control* (ZQC) and with it the use of *poka-yoke* devices (*poka* means inadvertent mistake and *yoke* means prevent). Shingo argued that the causes of defects lie in worker errors, and defects are the results of neglecting those errors. Detecting these mistakes before they become defects will lead to higher product quality. Shingo set about creating devices, usually immediate and visual, that allowed the operator to determine if he or she had made a mistake. An example often cited by Shingo relates to some work that he was doing in the Yamada Electric plant in 1961. Shingo recalled,

While visiting the Yamada Electric plant in 1961 I was told of a problem that the factory had with one of its products. Part of the product was a small switch with two push buttons supported by two springs. Occasionally, the worker assembling the switch would forget to insert a spring under each push-button. Sometimes the error would not be discovered until the unit reached a customer, and the factory would have to dispatch an engineer to the customer site to disassemble the switch, insert the missing spring, and re-assemble the switch. This problem of the missing spring was both costly and embarrassing. Management at the factory would warn the employees to pay more attention to their work, but despite everyone's best intentions, the missing spring problem would eventually re-appear. (Shingo, 1986, p. 43)

Shingo suggested a solution that became the first *poka-yoke* device:

In the old method, a worker began by taking two springs out of a large parts box and then assembling the switch. In the new approach, a small dish is placed in front of the parts box and the worker's first task is to take two springs out of the box and place them on the dish. Then the worker assembles the switch. If any spring remains on the dish, then the worker knows that he or she has forgotten to insert it. (Shingo, 1986, p. 44)

Many examples of *poka-yoke* devices appear in the literature. Indeed, L. J. Ricard (1987) describes an example of a *poka-yoke* device at General Motors:

[W]e have an operation which involves welding nuts into a sheet metal panel. These weld nuts will be used to attach parts to the car later in the process. When the panel is loaded by the operator, the weld nuts are fed automatically underneath the panel, the machine cycles, and the weld nuts are welded to the panel. You must remember these nuts are fed automatically and out of sight of the operator, so if the equipment jams or misfeeds and there is no part loaded, the machine will still cycle. Therefore, we have some probability of failure of the process. An error of this nature is sometimes not detected until we actually have the car welded together and are about to attach a part where there is not a nut for the bolt to fit into. This sometimes results in a major repair or rework activity ... to correct this problem, we simply drilled a hole through the electrode that holds the nut that is attached to the panel in the welding operation. We put a wire through the hole in the electrode, insulating it away from the electrode so as it passes through it will only make contact with the weld nut. Since the weld nut is metal, it conducts electricity and with the nut present, current will flow through, allowing the machine to complete its cycle. If a nut is not present, there will be no current flow. We try to control the process so that the machine will actually remain idle unless there is a nut in place.

Many studies of TPS have been undertaken, and some corporations have sought to implement their own approaches to the same concepts. In the transcript of a briefing to analysts that appears on the Eastman Kodak website,[1] Dan Carp, who was at the time CEO of Eastman Kodak, made the following statement:

In business theory it may be known as the Toyota management system right? Frankly making cars and making materials is two different things so you couldn't just drop that in. So the manufacturing teams have gone to school on some of these processes and they have been able to fundamentally change by product flow the way that we do business. It has led to lower inventories. In fact, customer satisfaction has gone up because we have been able to fulfill orders better and have perfect customer response. It is a pull system. It has allowed us to avoid capital. Our productivity in our plants is exceeding plan and our employees are more satisfied because at its very basic core, once the framework has been set, is that it allows the person on the line to make some key decisions about what we do.

One might be forgiven for thinking that this initiative by Eastman Kodak is an isolated case. On the contrary, Womack and Jones wrote a book entitled *The Machine That Changed the World* (Womack and Jones, 1991). In this book, they detail how companies can dramatically improve the performance of a company by introducing the concepts pioneered by Toyota. Womack and Jones coined the phrase *lean thinking*, which extends the thinking beyond the Toyota Production System put forward by Taiichi Ohno. Part of their rationale for introducing these concepts in their own way was the need to consider an approach for companies who already own manufacturing facilities. Clearly, a new or greenfield site would be ideally suited to the Ohno concepts. An existing manufacturing process is another matter entirely. In their more recent book *Lean Thinking: Banish Waste and Create Wealth in Your Corporation* (Womack and Jones, 1996), they try to answer the questions that you might want to pose, such as "What is the next step once you become a Toyota-like company?" Much of the recent literature following on from the Womack and Jones books uses the term *lean* in one form or another. Womack and Jones also provide the interested reader with a means of implementing lean thinking into a mature manufacturing site.

They (Womack and Jones) provide the following definition of *lean thinking*:

> Fortunately, there is a powerful antidote to muda (waste)—lean thinking. It provides a way to specify value, line up value-creating actions in the best sequence, conduct these activities without interruption whenever someone requests them, and perform them more and more effectively. In short lean thinking is **lean** because it provides a way to do more and more with less and less—less human effort, less equipment, less time and less space—while coming closer and closer to providing customers with what they want. (1996, p. 15)

Their basic premise is that a customer will be more disposed to "pulling" goods through a manufacturing process if the goods have high *value* to that customer. Having determined that the goods in question have value (even though the value may be hard to quantify in absolute terms), one then looks at the whole process from concept and design, through development and manufacture, looking at the *value stream*. This exercise looks at all of the activities associated with producing the item in question from ordering the raw materials to producing final goods. This drives to the issue of the *lean enterprise*, which, once identified, can be examined to uncover *all* sources of waste, which usually fit into two categories:

- Unavoidable waste (termed *Type 1 muda*)
- Avoidable waste (termed *Type 2 muda*)

Having specified the value and mapped the value stream using the lean enterprise, Type 2 muda can be removed. The next concept to consider is the *flow* of materials through the remaining steps, be they ordering, research and development, or manufacturing and delivery. Just as we defined *kaizen* earlier to be continuous incremental improvement, so there is also a term for radical improvement, or *kaikatu*. *Kaikatu* events are used to describe radical change of the process, the objective for which is a major improvement in flow.

3.2.1 Mistake Proofing (*Poka-Yoke*)

It's hard if not impossible to determine if the mistake-proofing concept within TPS drove designers to incorporate some of the concepts into modern electronics or indeed if it would have happened on its own. In any event, there are many examples in the electronics and electrical goods industries where mistake proofing is a way of life. Figures 3.6, 3.7, and 3.8 are three simple examples of mistake proofing in the domestic environment that should help to demonstrate the principle.

In Figure 3.6, showing a random access memory (RAM) unit from a personal computer (PC), electrical contacts are at the bottom of the unit and have a notch deliberately separating some of the electrical contacts. This notch is manufactured to fit a node on the docking port. The notch is not in the middle of the block. There is therefore only one position for which this unit can be fitted.

Figure 3.7 is a UK electrical mains socket where plugs can only fit the socket if they have three pins of the correct orientation. Additionally, there is only one way that they will fit. It is therefore not possible to incorrectly fit a plug to a socket, and so the electrical contact between the mains ring circuit and the appliance has been mistake proofed.

Finally, Figure 3.8 shows three rows of pins in slightly different positions surrounded by a protective locator sleeve. This sleeve is not a true

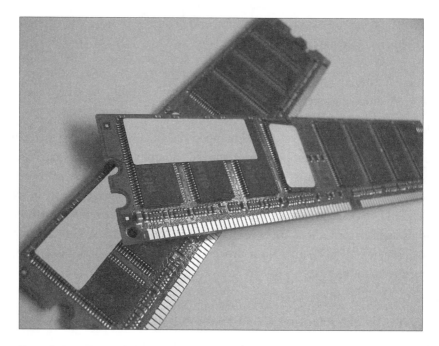

FIGURE 3.6
Random access memory (RAM).

oblong but has a shape that will only match the female connector in a single orientation.

These designs have not happened by accident. Careful thought has gone into their design so that there is no possibility of error. This is the essence of mistake proofing or *poka-yoke*. Industrial software programmers have used mistake-proofing techniques for many years in software design where there is a need to create interlocks that prevent excessive pressure buildup (during a distillation, for example) or interlocks to prevent chemical discharges either into the drainage system or vented to the atmosphere.

FIGURE 3.7
Socket.

These techniques, however, were not always as rigorously applied to the mechanical equivalents. For example, car manufacturers used to employ operators to fit nuts, bolts, washers, and springs to various areas of dashboards on car assembly lines. Many years ago, the

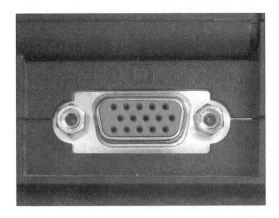

FIGURE 3.8
Visual display unit (VDU) socket.

operator had to remember how many of each of the washers, nuts, bolts, and springs needed to be fitted to the unit and to also take the required number from larger containers. The TPS mistake-proofing concepts provided the operator with the exact number of washers, nuts, bolts, and springs. The operator then has to fit all of the parts prior to moving on to the next unit without having to remember if they had all been fitted. Any parts left over were obvious and provided the operator with the opportunity to check his or her own work.

This is a trivial but real example of how a simple change to an operating procedure can help to make a large difference to the outgoing product quality. The above example is relatively old and now might seem trivial. At the time, however, it was revolutionary because mistake-proofing concepts were in their infancy.

3.2.2 Stop the Process (*Jidoka*)

Of all of the various elements of the Toyota Production System, *jidoka* most closely resembles Toyoda Sakichi's automatic loom. As you may recall, this loom stopped spinning and weaving when one of the threads broke. As a previous employee of the Toyoda Spinning and Weaving Company, Taiichi Ohno would have been only too aware of the need to develop the principle of stopping a process before defective products were produced.

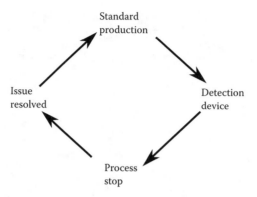

FIGURE 3.9
The *jidoka* cycle.

It is an interesting issue since, if taken to its logical conclusion, there will be no need for a quality control or inspection department as there will be no defects to inspect! Additionally, all products leaving the assembly line at Toyota are defect free and can therefore be shipped directly to the customer without the need for on-site storage. You may recall that inventory is one of the seven orders of waste identified in TPS.

The idea behind *jidoka* is simple to understand but may be very difficult to enable (Figure 3.9). For example, an implementation might involve installing a simple trip switch that detects a change in line speed, orientation, tension, pressure, temperature, and so on. This trip switch then sends a signal to the process line computer, and the line stops. Ideally, the detector should not be an operator as operators become tired, can be distracted, may be momentarily involved in other duties, or the like. There is no reason why an operator might also be able to stop the process, but there should also be a mechanical means of ensuring that the process is always functioning correctly. Restarting the process then takes place once the issue that caused the defect has been fixed.

Figure 3.3 showed how many facets there are to full TPS implementation. Clearly, it will take months or even years to implement all of the system. There is also a need to invest in a cultural change within the organization and also to encourage the workers to change their working attitudes and practices. All too often, the so-called soft skills involving behavioral change are either totally neglected or given lip service. A report on the Toyota website[2] discusses some of these aspects through the statements made below:

- A 'pull' system asks workers to use their heads
  - This is a production system driven by actual consumption and controlled by synchronized replenishment signals. Nothing is produced by the upstream supplier until the downstream customer signals a need. This system allows for continuous flow whereby products flow like water through a pipe. This makes the production cycle predictable and repeatable and maximizes the throughput of finished product
- In TPS, the T also stands for 'Thinking'
- Ask yourself 'Why?' five times
  - This technique is designed to get back to the root cause of a problem
- Train people to follow rules and standards as if second nature
  - Standardized procedures, practices and activities are executed consistently and in regular time intervals to ensure that the 5 S's (sort, set in order, shine, standardize and sustain) are guaranteed
- Develop people who can come up with unique ideas

Some other aspects to TPS are of equal importance:

- Cut lead time; indeed, cut out all the bits that don't add value.
- Deal with defects only when they occur, and the number of staff you need will drop.
- Find where a part is made cheaply, and use that price as a benchmark.

From the very outset, Toyoda Kiichiro outlined his philosophy to which he always adhered. They became known as *Toyotaism*. He placed the following conditions on the automotive business:

- To provide cars for the general public
- To perfect the passenger car industry
- To make reasonably priced cars
- To recognize the importance of sales in manufacturing
- To establish the basic material industry

Kiichiro published an article in September 1936 entitled "Toyota to the Present," during which he further elaborated on Toyotaism. Ohno quoted passages from this article in his book, some of which prove insightful:

[P]roblems differed from those of weaving machines, however, and we realised that the new business would be difficult to create. So, for three years we managed the business under the guise of a hobby. But the unexpected lapse in automobile manufacturing forced us to take a business attitude—not a hobbyist's. The business now involves an obligation to the country. Whether we like it or not, we have to make it work as soon as possible. (Quoted in Ohno, 1988, pp. 80–82)

He went on to describe what preparations had been undertaken during the three years that his fledgling business was known as a "hobby." Much of his discussion concerns the need to ensure that the basic materials of steel and parts were fabricated with the correct quality.

These data show that the expansion of the Toyota Motor Company (Figure 3.10) was hand in hand with the evolution of TPS. Financial data for 2009 show that the Toyota Motor Company is now the largest motor manufacturing company in the world. There must be something to this system!

So, in summary, this waste reduction system has been proven to work and is based on ten basic "rules" (see, e.g., Liker, 2003):

- Abolish local optimization.
- Create a culture of continuous improvement.
- Do it right the first time.
- Eliminate waste.
- Empower workers.
- Maximize flow.
- Meet customer requirements.
- Minimize inventory.
- Partner with suppliers.
- Pull from demand.

These seemingly simple rules have been implemented in many industries, including construction, customer service, finance, health care, and logistics. The application of the rules may change slightly from one industry to the next, but the underlying principles have stood the test of time.

Year	Event
1930	Kiichiro Toyoda started research into gasoline-powered engines.
1933	Automobile Department established in Toyoda Automatic Loom Works Ltd.
1935	Hinode Motors (currently Aichi Toyota) started operations.
1936	Toyota's logo established.
1937	Toyota Motor Co., Ltd. established.
1938	Koromo Plant (currently Honsha Plant) started operations.
1940	Toyoda Seiko, Ltd. (currently Aichi Steel Works, Ltd.) established.
1940	Toyoda Physical and Chemical Research Institute established.
1941	Toyota Machine Works Co., Ltd. established.
1943	Tokai Hikoki Co., Ltd. (currently Aisin Seiki Co., Ltd.) established.
1945	Toyota Shatai Kogyo Co., Ltd. (currently Toyota Auto Body Co., Ltd.) established.
1946	Kanto Electric Auto Manufacturing, Ltd. (currently Kanto Auto Works, Ltd.) established.
1948	Nisshin Tsusho Co., Ltd. (currently Toyota Tsusho Corporation) established.
1949	Nagoya Rubber Co., Ltd. (currently Toyoda Gosei Co., Ltd.) established.
1949	Nippondenso Co., Ltd. (currently Denso Corporation) established.
1950	Financial crisis, labor dispute, and voluntary retirement. Toyota Motor Sales Co., Ltd. established.
1950	Minsei Spinning Co., Ltd. (currently Toyoda Boshoku Corporation) established.
1953	Towa Real Estate Co., Ltd. established.
1961	Haruhi Plant (currently Haruhi Center) completed.
1965	The Deming Prize awarded. Kamigo Plant started operations.
1966	Takaoka Plant started operations.
1968	Miyoshi Plant started operations.
1975	Shimoyama Plant started operations. Prefabricated housing business started.
1978	Kinuura Plant started operations.
1979	Tahara Plant started operations.

FIGURE 3.10
Toyota timeline: The first 50 years.

NOTES

1. http://www.kodak.com.
2. http://www2.toyota.co.jp/en/index_company.html.

REFERENCES

Liker, J. K., ed. 1997. *Becoming Lean: Inside Stories of U.S. Manufacturers.* New York: Productivity Press. ISBN-10 1-563-27173-7.

Liker, J. 2003. *The Toyota Way: 14 Management Principles from the World's Greatest Manufacturer*. Maidenhead, UK: McGraw-Hill. ISBN-10 0-071-39231-9.

Ohno, T. 1988. *Toyota Production System: Beyond Large-Scale Production*. New York: Productivity Press. ISBN-10 0-915-29914-3.

Ohno, T. 1989. Bringing Wisdom to the Factory. In S. Shingo, *A Study of the Toyota Production System*. New York: Productivity Press. ISBN-10 0-915-29917-8.

Ricard, L. J. 1987. GM's Just-in-Time Operating Philosophy. In Shetty, Y. K., and V. M. Buehler, *Quality, Productivity and Innovation*. New York: Elsevier Science Publishing. ISBN-13 978-0444011947.

Shingo, S. 1985. *A Revolution in Manufacturing: The SMED System*, trans. A. Dillon. New York: Productivity Press. ISBN-10 0-915-29903-8.

Shingo, S. 1986. *Zero Quality Control: Source Inspection and the Poka-yoke System*. New York: Productivity Press. ISBN-10 0-915-29907-0.

Shingo, S. 1989. *A Study of the Toyota Production System from an Industrial Engineering Viewpoint (Produce What Is Needed, When It's Needed)*. New York: Productivity Press. ISBN-10 0-915-29917-8.

Wild, R. 1989. *Production and Operations Management*. London: Cassell Educational Limited. ISBN-10 0-304-31592-3.

Womack, J. P., and D. T. Jones. 1991. *The Machine That Changed the World: The Story of Lean Production*. London: Harper Perennial. ISBN-10 0-060-97417-6.

Womack, J. P., and D. T. Jones. 1996. *Lean Thinking: Banish Waste and Create Wealth in Your Corporation*. London: Simon and Schuster. ISBN-10 0-684-81035-2.

4

Global Motor Manufacturing in the Credit Crunch of 2008–2009

Before we have a look at what the motor manufacturers have been doing, let's have a look at TPS as applied to inventory reduction.

Throughout his working life, Ohno was really only interested in one issue, that of the time taken from a customer placing an order to the company receiving the money for the goods sold (Ohno, 1988; and see Figure 4.1). Delays in the time taken from customer order to cash received result in waste to the company. Minimizing time to cash cannot be fully achieved without first addressing all waste issues. Most, if not all, internet-based companies have realized the value of time to cash and exploit it to full advantage. They do this by charging a credit card when the order for goods is placed. In this way, the company has your money before you have the goods. The really smart internet companies buy the components to manufacture the product that they have just sold *after* they have taken your money. They use the customer's money to fund their own business, often delaying payment for their own suppliers for as long as possible.

Of the seven orders of waste identified by Ohno, the most damaging to any company is that of overproduction. Ohno made many comments concerning overproduction, including the following:

> [C]onsider the waste of overproduction, for example. It is not an exaggeration to say that in a low-growth period such waste is a crime against society more than a business loss. Eliminating waste must be a business's first objective.... (1988, p. 129)
>
> When thinking about overproduction, I often tell the story of the tortoise and the hare.

FIGURE 4.1
Time to cash.

In a plant where required numbers actually dictate production, I like to point out that the slower but consistent tortoise causes less waste and is much more desirable than the speedy hare who races ahead and then stops occasionally to doze. The Toyota Production System can be realised only when all the workers become tortoises. (1988, p. 62)

Other companies that have implemented TPS techniques have also found that overproduction in their hands is also the worst of the seven orders of waste. But why should this be the case? As we have seen earlier, many TPS techniques are basically concerned with minimum inventory, motion, transportation, and so on.

The basic answer is that finished products have had all of the time and money invested in their manufacture that a company can provide. Additionally, an invoice for the raw materials may have been received, workers will have been paid, factory overheads may have been paid, and so on.

This concept also applies to nonmanufacturing industries. Consider a consultant or an author writing material that is never used. It could be PowerPoint slides that are never shown, a brochure describing a workshop that was never given, or a chapter of a book that was never published. There are many examples.

It is tempting to make the following blanket statement:

If overproduction is the worst form of waste, then the simple answer is to reduce overproduction.

Actually, it is easy to make the statement but very difficult to implement a change to the status quo within a company for a whole host of reasons. It is rather like trying to jump onto or jump from a moving carousel and hoping to survive the transition from stationary to moving or vice versa.

It took Ohno and others almost thirty years to develop and implement these techniques. In some cases, the development phases took many years. However, they are now available—so there really is no excuse!

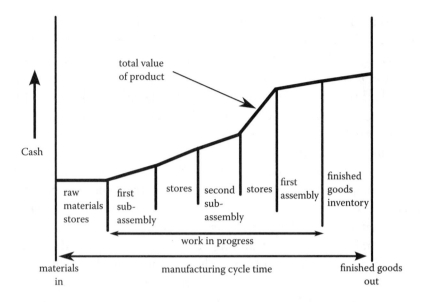

FIGURE 4.2
The cost of inventory.

If there is one driver that will crystallize the mind, it is that of inventory reduction. Inventory and overproduction are intimately linked.

Any delay in the time to cash creates inventory somewhere in the system (see Figure 4.2). In Figure 4.2, there is repeated inventory of raw materials; the first subassembly, the second subassembly, and the final product. For the chemical industry, this relates to storing raw materials, intermediates, and final product. The motor-manufacturing industry is more likely to store finished goods as inventory because they often manufacture cars using assembly lines. There are a few examples of companies producing bespoke vehicles for the high end of the market—Bugatti, for example; however, these manufacturers are few and far between.

The 1970s was certainly a fashionable time for company chairmen to hold warehouse opening ceremonies. Not only were these opening ceremonies high-profile events, but also they resulted in a financial drain on the company from many perspectives. Warehouses or silos

- are expensive to run and maintain,
- lengthen the time to cash, and
- can result in out-of-date or out-of-fashion product.

Most modern electronic goods, such as mobile phones, would definitely be superseded if they were stored for any length of time in a warehouse as the colors, ringtones, and other features preprogrammed in the product would soon become outdated. As other examples, chemicals may start to degrade through a number of mechanisms such as heat, light, moisture, and oxygen; motor vehicles are susceptible to rust; and on-board stereo systems would become outdated.

Assuming defect-free product during manufacture (i.e., that all of the concepts discussed in Chapters 2 and 3 have been implemented), inventory buildup is most likely in batch processing. These concepts have been used in most manufacturing companies for many years (i.e., orders are accumulated by the planning departments until a predetermined minimum number is reached). Planning departments then place the product in the plan, and goods are subsequently manufactured. The finished goods are then placed in a warehouse from which orders are picked to supply the customers. Figure 4.3 outlines a simplified scenario, with four customers ordering product that is red, green, yellow, and blue.

Fulfilling customer orders is a complex issue. In Figure 4.3, the blue order was the fourth order to be placed. With a batch-manufacturing process, however, the blue order cannot be satisfied until all of the red,

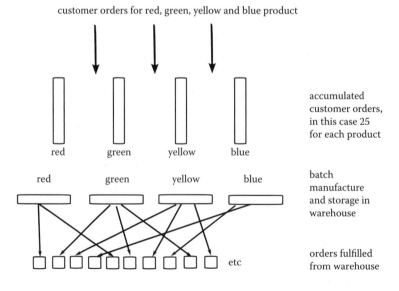

customer orders for red, green, yellow and blue product

red   green   yellow   blue

accumulated customer orders, in this case 25 for each product

red   green   yellow   blue

batch manufacture and storage in warehouse

etc

orders fulfilled from warehouse

FIGURE 4.3
Fulfilling customer orders.

red	green	yellow	blue	red	green	yellow	blue	etc.	orders
red	green	yellow	blue	red	green	yellow	blue	etc.	manufacturing
red	green	yellow	blue	red	green	yellow	blue	etc.	product available

FIGURE 4.4
Product leveling.

green, yellow, and blue products have been manufactured and sent to the warehouse. Depending upon the rate at which the orders come in, and the speed of manufacture, warehousing, and product distribution, it may be days or weeks before the blue order can be fulfilled.

The TPS concept of product leveling uses the alternative scenario of manufacturing the orders as they come (see Figure 4.4). Product leveling requires that a product is manufactured *only* if there is a customer order. In other words, if only four orders come in for the day, then only four products are made. The product is shipped to the customer from the manufacturing site and not placed in any intermediate warehouse.

Of course, life is not this simple; additionally, there is also the issue of customer requirements, which may change at short notice. In Chapter 3 I mentioned the 1970's fuel price increases. The Toyota Motor Company had no stock of finished vehicles, and when the need arose to manufacture cars and vans with lower engine capacity they responded the next day. Some other car manufacturers had old stock, not all of which could be sold for the original price. They also had to change their supply chain, which in some cases took months and not minutes.

Many issues face manufacturing divisions prior to the complete implementation of product leveling. For example, the product technologists, engineering community, and production staff need to do the following:

- Start the process with minimum waste.
- Stop the process with minimum waste.
- Ensure minimum downtime between products.
- Ensure product scheduling.
- Ensure machine setup.
- Correctly obtain relevant paperwork.
- Obtain correct materials flow for oncoming grade.
- Remove completed product from production area.

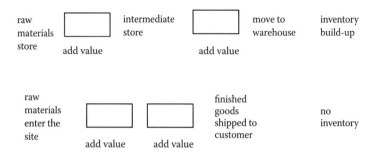

FIGURE 4.5
Manufacturing scenarios with and without inventory.

There are other issues, such as product change during a shift change, which need effective and timely communication, among other things. While these issues may seem daunting, if not insurmountable, these issues have been overcome by many companies implementing TPS.

Storage of intermediates, be they chemical, electrical, or mechanical, is expensive, but what is the alternative? The first example in Figure 4.5 shows only two manufacturing steps, but there could equally be a multistage process. The second scenario shows the finished goods being shipped directly to the customer. The logical outcome of this is that the only goods manufactured under the second scenario will be those that the customer has specifically ordered. A *pull system* is also known as *just-in-time* as the goods are manufactured to order and not for inventory.

This scenario is therefore a pull system where the materials that flow through the manufacturing site are driven by customer orders. In this case, raw materials entering the site do so only when there is a customer need for product. Clearly this is an extreme, and there will need to be some inventory; however, what little there is *must* be justified.

During the development of TPS, Ohno realized the importance of the American supermarket system of replenishing stocks of materials for supply internally as well as to the customer. Ohno commented,

> From the supermarket we got the idea of viewing the earlier process in a production line as a kind of store. The latter process (customer) goes to the earlier process (supermarket) to acquire the required parts (commodities) at the time and in the quantities needed. The earlier process immediately produces the quantity just taken … in 1953 we actually applied the system to our machine shop in the main plant. (1988, p. 26)

The "supermarket" thus became an integral part of TPS; however, it carries inventory only when it is absolutely necessary. Some of the recognized instances where supermarkets are permitted include the following:

- A physical distance between supplier and user that prohibits one-piece flow.
- A work center supplying more than one product line.
- Supplier product reliability. Supplier issues need to be dealt with by ensuring that they are part of the overall process as partners.

One-piece flow is the term used to define the manufacture of product one piece or item at a time.

Supermarkets are set up with the following guidelines:

- Materials are withdrawn and replenished in small quantities.
- Materials enter and leave the supermarket using the first-in, first-out methodology.
- Only minimum inventory of any one item is allowed.
- Supermarkets are kept tidy.

The layout of a supermarket that serves several production lines is important (see Figure 4.6). Remember that the term *supermarket* is used differently here compared with high street shops. The example in Figure 4.6 considers a supermarket with four items, supplying materials to four production lines. In this case, material C from the supermarket only supplies intermediates to production line 1, shown as just one arrow leaving "bin" C. It might be the case that the items in the supermarket should be reorganized so that item C appears at the top of the diagram as that would

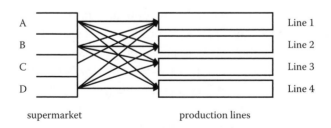

FIGURE 4.6
Supermarket layout.

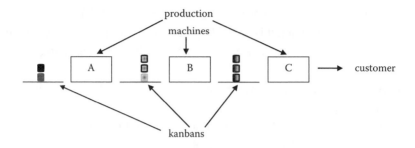

FIGURE 4.7
Kanban principles.

involve less motion than the current order. You also need to consider the frequency at which items are withdrawn from the supermarket shelves.

The process of good communication between the customer, manufacturing departments, and raw materials supply can be achieved using a number of techniques. In practice, the concept of a *kanban* was developed at the Toyota Motor Company. *Kanbans* are simple but nevertheless very powerful means of communicating upstream requirements.

4.1 WHAT EXACTLY IS A *KANBAN*— WHAT DOES IT LOOK LIKE?

Kanbans can be boxes, cards, flags, tubes, or even a square painted on the floor. Figure 4.7 is an attempt to demonstrate the use of a *kanban* for the three-stage process.

At the start of the cycle, all of the *kanbans* are full (shown as three different types of square) and so production has stopped (see Figure 4.7). A customer order is then received. Production machine C takes the intermediate from its *kanban* and creates the product (see Figure 4.8). Production machine B notices that there is an empty *kanban* and manufactures a replacement which is placed in the relevant *kanban* (see Figure 4.9). Production unit A then notices that a *kanban* is empty and makes one of its products (see Figure 4.10). The final step in the process recognizes that the raw material *kanban* is empty, and a new batch of raw material is purchased so that the "factory" is once again ready for production (see Figure 4.11).

FIGURE 4.8
Kanban between C and B needs filling.

FIGURE 4.9
Kanban between B and A needs filling.

FIGURE 4.10
Kanban A needs filling.

FIGURE 4.11
Full kanban.

Each *kanban* in the sequence

- cannot exist separately.
- minimizes material handling.
- quickly identifies quality issues.
- reacts to the other *kanbans*.
- simplifies the ordering process.
- works in combination with the other *kanbans*.

In the above example, one part was moved through the "factory" so as to demonstrate the principle. In practice, there may be many *kanbans* supplying intermediates for one of the manufacturing units.

Good communication is a *must* as everyone needs to know what is expected of them. Standardized work is therefore a key of a successful *kanban*. Let's have a quick look at multiple *kanbans*, which will after all be closer to a real production line (see Figure 4.12). In this example, there are four *kanbans* serviced by one department. Assuming that

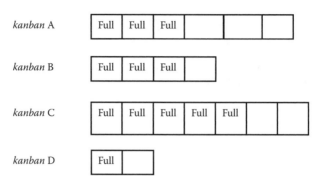

FIGURE 4.12
Various *kanbans*.

product leveling has been implemented, there should be a steady flow of material into and out from the *kanbans*. The empty boxes denote work that is needed from the supplier process to fill the *kanbans*. Under these circumstances, which *kanban* should be filled first? There is some help in the time slot board. This visual control allows the operators to keep track of when certain operations should be started (see Figure 4.13).

In this particular case, the time slot board has filled up as the *kanbans* have been emptied. A different day might result in a different ordering as the customer production lines change their products in line with customer orders. All of these communications issues are addressed in standardized work; after all, it is just a case of everything being in the right place at the right time.

Let's just switch gears a little and discuss motion in more detail. What we are really looking at is the frequency of movement and distance traveled.

07:00–10:00	Fill *kanban* C
10:00–12:30	Fill *kanban* D
13:00–15:00	Fill *kanban* A
15:00–19:00	Fill *kanban* B

FIGURE 4.13
Time slot board.

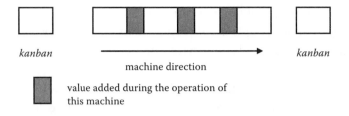

kanban   kanban

machine direction

value added during the operation of
this machine

FIGURE 4.14
Manually operated production line.

While these issues are inevitable in a factory or even a department, you should try as far as possible to minimize the length of time workers move, especially if there is no added value to the product.

Materials supply to a production unit comprise just one consideration; they may not be the primary source of non-value-added waste activity. Consider Figure 4.14, which depicts an operator-driven batch process. This classic layout for an operator-driven batch production machine has been shown with two *kanbans*: a supply and completed work. This is not the final stage in the overall production sequence, as the latter *kanban* becomes the source *kanban* for the next operation. The gray areas represent value-added activities. There is no scale on the diagram, and so there is no idea of the distances walked by the operator between the two *kanbans*, nor between the value-added areas of the workstation. This is deliberate.

Figure 4.15 proposes a change to the layout for this machine to minimize operator movements. These types of U-shaped work cells are usually designed to work anticlockwise. There are many reasons for this, some being the following:

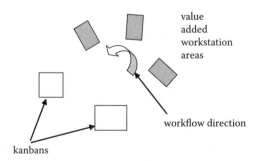

value
added
workstation
areas

workflow direction

kanbans

FIGURE 4.15
U-shaped work cell.

- Over 85 percent of the population is right handed.
- Right-handed people usually have a dominant right eye.
- Right-handed people often have a dominant right leg.

An operator using the layout in the machine outlined in Figure 4.15 will walk a shorter distance and has the opportunity to be more productive. Layout considerations have four basic elements. The optimum layout will

- ensure that the operator movements are not hindered by material movements.
- ensure that operator flexibility is more important than process order.
- minimize the number of operators.
- minimize operator movements.

This simple workplace redesign will not be possible with kettles or other machinery too costly to redesign, nor is it appropriate if the unit production length requires a linear process. A linear example would be a glass manufacturer using floating glass techniques. In the main, these types of products are the exception and not the rule.

One of the TPS metrics often used to determine the available time an operator can work is called *takt* time. Various definitions exist, including the following:

- The available work time per shift
- The customer demand rate per shift
- The frequency at which a product should be produced based on the rate of customer sales
- The time required between successive units of end product existing as a production operation

Ohno described *takt* time as follows:

[T]he length of time, in minutes and seconds, it takes to make one piece of the product. It must be calculated in reverse from the number of pieces to be produced. [Takt] is obtained by dividing the operable time per day by the required number per day (pieces). Operable time is the length of time that production can be carried out per day. (1988, p. 60)

$$\frac{\text{available time per shift}}{\text{customer demand rate per shift}}$$

FIGURE 4.16
Takt time calculation.

However defined, the calculation of *takt* time is always described as in Figure 4.16, unless the units are changed from shift to minutes, hours, days, or some other unit of time. *Takt* time should be synchronized across a workflow. If this does not happen, there will be a buildup of inventory in one of the *kanbans* or the process will stop because there is no inventory in the *kanban*. In either event, the *kanbans* will signal a problem.

If there is an issue of *takt* time synchronization between different products, there are a number of potential solutions. In all of the examples provided in Figures 4.17, 4.18, and 4.19, linear production lines have been drawn and not U-shaped work cells. This is for convenience only.

In these figures, R, G, B, and Y denote four products.

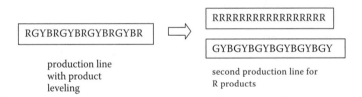

RGYBRGYBRGYBRGYBR

production line
with product
leveling

RRRRRRRRRRRRRRRR

GYBGYBGYBGYBGYBGY

second production line for
R products

FIGURE 4.17
Resolving takt time imbalance – example 1.

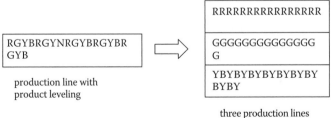

RGYBRGYNRGYBRGYBR
GYB

production line with
product leveling

RRRRRRRRRRRRRRRR

GGGGGGGGGGGGGGG
G

YBYBYBYBYBYBYBY
BYBY

three production lines
added to compensate for
takt time imbalance

FIGURE 4.18
Resolving takt time imbalance – example 2.

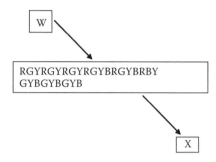

FIGURE 4.19
Feeder or offshoot cells.

Create multiple production lines, for example Figures 4.17 and 4.18. In the extreme, several production lines may be needed (see Figure 4.18). Create small feeder cells, sometimes called *offshoot cells* (see Figure 4.19).

Where equipment is expensive to duplicate, subassemblies can be built offline and added to the main production line for further work. Similarly, partially assembled product can be removed. Under these circumstances, components and completed subassemblies could be totally unrelated to any of the other products. The incoming raw material and outgoing subassembly are shown as W and X in Figure 4.19.

Add additional work to the underutilized cell (see Figure 4.20). Figure 4.20 represents an overall production process which consists of three separate machines. In this example, let's assume that the middle machine has some spare capacity compared with the other two machines. An additional product, represented by P, is added to the production schedule for this middle machine. This adds value to product P at the same time as balancing out the *takt* time between the three production machines.

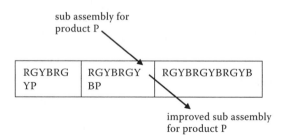

FIGURE 4.20
Method for balancing takt time.

There are some basic guidelines for equipment, cell design, and redesign proven to reduce waste. These rules have been generated as a result of studying many industries. While every effort should be made to move toward these suggestions, some sectors of the chemical industry will struggle to implement most of these suggestions. Nevertheless, the full list is provided for completeness. It is as follows:

- Ensure that as far as possible, equipment should be completely self-contained, be easy to move, and connect to relevant services (water, electricity, compressed air, etc.).
- Machines are built with setup in mind.
- They are easy to maintain.
- They are easy to operate with visual controls.
- They are easy to reconfigure.
- Equipment should be as close to "load-load" as possible, that is, as automatic as possible where operators only undertake value-added steps so that the operator and machines perform the task together.
- Equipment should be as narrow as possible. Ideally, the width of the machine should be just wider than the widest product.
- Equipment should be narrow and deep, operated from the front and supported from the back.
- Equipment should be safe, ergonomic for the operator, and easy to clean.
- Machines and equipment should have high-process capability and include appropriate failsafes and mistake-proofing capability.
- Product should move through the equipment one item at a time such that it is not possible to create batches.
- The movement of product through the machine should be balanced to *takt* time.

Some suggested best practices for setting up and running work cells that follow the above guidelines which are more concerned with engineering best practices and not applicable to the majority of equipment found in chemical plants.

There are a lot of best practices in this chapter that we can test against the motor manufacturers. Such a test will help us try to understand the potential management of the industry. It could be the case that each motor manufacturer has a senior management team committed to best practices and that this quick look at the industry will reveal nothing. On the other hand, if there were problems going into this recession that are not addressed before the economic upswing, the issues and concerns will surface. This will be a major problem if we as taxpayers are propping up businesses that could benefit from understanding TPS in more detail. The other issue, to be explored in the Chapter 5, is the conditions associated with the government bailouts which will need to be addressed as well as the issues of waste reduction activities.

Let's start with some pictures that have appeared in the UK national newspapers, particularly the *Times*. The pictures that I have in mind show countless numbers of parked vehicles. Some locations of these cars are Sheerness docks, Nissan's Sunderland manufacturing site, Southampton docks, a disused airfield near Oxford, the Avonmouth docks, and several other pictures. Between them, they must show tens of thousands of new vehicles.

First of all, let's look at the possibility that these different manufacturers are each running pull factories (i.e., producing vehicles to match orders and not simply building inventory). Each vehicle will have

- been parked by a driver,
- some petrol in its tank,
- keys that need to be documented and stored,
- inevitable rust spots that will be produced over time,
- become outdated as fashions change,
- tied up company investment capital,
- depreciated in value with time,
- required round-the-clock security, and
- taken up space that could be rented for other purposes.

Additionally, the site will need to be maintained. The vehicles will need some form of inventory management so that if some were to be sold, the paperwork for the sale could be transacted and so on. This is a real and visible issue. Where were the performance and productivity data that led to this level of overproduction? There seems to have been a disconnect between customer orders and units produced. This philosophy of maximizing volume does indeed keep workers and machines occupied but also consumes raw materials. If there is a customer, ensure that the product is delivered as quickly as possible and that payment for goods received

	Total	
Production in Japan		
Passenger cars	194,560	(−52.6)
Trucks and buses	35,474	(−41.4)
Total	230,034	(−51.1)
Sales in Japan		
Passenger cars	183,797	(−26.6)
Trucks and buses	30,197	(−30.2)
Total	213,994	(−27.1)
Exports		
Passenger cars	74,333	(−69.2)
Trucks and buses	11,662	(−62.6)
Total	85,995	(−68.5)
Overseas production	242,002	(−40.3)
Worldwide production	472,036	(−46.1)

FIGURE 4.21
Toyota production.

is prompt. So the likelihood is that some of these vehicles were produced to keep the factories open and not to fulfill customer orders. There is no value to inventory for which there is no customer.

As we saw in Chapter 3, the Toyota Motor Company only produces vehicles for which they have customers so there are no pictures of vast numbers of Toyota cars waiting to be sold. However, Toyota is still suffering from the effects of the recession! Figure 4.21 shows some data from their website for their March 2009 results.[1]

The consolidated data in Figure 4.21 show a worldwide production of vehicles at 46.1 percent of their normal production volumes. This decrease in overseas production is the first for seventeen years and shows the strength of the company. The Toyota management team is dealing with the problem as they have with all other changes in the past. They do this by reducing production as there are fewer sales. They are still operating a pull system of manufacturing, so no orders equates to no production. Chapter 5 will concentrate on the other vehicle manufacturers and the effects of the recession on their business. The product recalls in the spring of 2010 have only added to Toyota's problems.

NOTE

1. http://www.toyota.com.

REFERENCE

Ohno, T. 1988. *Toyota Production System: Beyond Large-Scale Production.* New York: Productivity Press. ISBN-10 0-915-29914-3.

5

The Impact of Global Attitudes and Bailout Conditions

In times of a crisis or serious discussion or concern, a former manager of mine used to describe the ensuing fallout as "a lot of churn." Without a doubt, there is a lot of churn in the financial markets at the moment! History (as well as a few authors and academics) will write the whole story in detail. Our concern here is the impact of the credit crunch, subprime, and recession issues of 2008–2009 on the uptake and progression of lean thinking concepts in businesses—at least in the Western countries.

Many small and medium enterprises have simply ceased to trade because they often relied on credit to keep their cash flows positive. As credit from the banks and financial institutions evaporated, so did financial support for their businesses. Unfortunately, there is little that can be done for them using lean thinking concepts, unless their owners can see a way of restarting their existing businesses or starting up new ventures. Attention to inventory control and waste in general will help set the tone for any (re-) emerging companies.

The finance sector itself has suffered the most churn for it was some members of these institutes that caused the problems in the first place. I have often thought of, and indeed tried to estimate, the number of misguided finance "experts" who caused the 2008 subprime debacle. It could be as low as 200 or as high as 5,000. It was certainly a minority of workers and board members who between them wreaked havoc with the world economy, the effects from which will be felt for many years to come. Lines of credit will slowly be extended, and once again businesses around the globe will slowly try to establish normality, but it will take time.

The finance sector itself embraced lean thinking concepts in the early part of 2000 and continued their efforts until the subprime churn in 2009. Unfortunately, events have overtaken them. What was once a stable working environment where well-constructed and reasoned ideas were accepted for implementation will, for a while at least, simply fall by the wayside. As the finance industry suffers temporary collapse, there will be widespread sector unemployment, and the culture of continuous improvement will, for a short time at least, cease to exist. Indeed, we have already seen the collapse of AIG and Lehman Brothers, and have seen many banks receive government aid. It may take some of these banks years to recover, and only time will tell if the future finance institute leaders of those will have the strength of character to build *lean* concepts into their reconstruction plans. They have certainly returned to some of their old habits (a high-bonus culture, for example); let's hope that waste reduction is also part of their long-term strategy.

Of one thing there will be no doubt: businesses and individuals will pay for this crisis one way or another, either through higher taxes so that their respective governments will slowly recover their finances or through higher interest rates from lending houses as they try to recover their positions in the relevant stock markets. So the effects on businesses will reverberate for many years to come.

But what of the motor industry, which might be considered to be an innocent bystander in the collapse of the financial markets? Global vehicle demand was down 21 percent in the first quarter of 2009 compared with the same quarter in 2008. If you look at just the top four manufacturers,[1] their net income indicates some issues (see Figure 5.1). A quick glance at

	% Loss, First Quarter 2009 versus First Quarter 2008
Toyota Motor Company	21.9
GM	*
Daimler	25
Ford	#

* GM reported a net loss of $6 billion for first quarter 2009 compared with first quarter 2008; however, they also reported a loss for the previous year of $5.8 billion (1 billion = 1,000,000,000).

\# After-tax results (in millions): $(1,427) for 2009, $(1,497) for 2008.

FIGURE 5.1
Effect of the recession.

the figures suggests that Toyota is tracking global demand, and Daimler is suffering more than might be expected.

One might think that all of the global top ten motor manufacturers would report losses of 20–25 percent for this period. Not so, however, at the Volkswagen Group, as their website posts the following:[2]

Volkswagen Brand Deliveries Increase by 1.3 Percent

The Volkswagen Group has again developed noticeably better than the global automotive market. In an overall market down by approximately 20 percent, the Group delivered 541,600 vehicles (April 2008: 568,100; −4.7 percent). 'With our Group's attractive model range we have largely been able to buck the trend and are on target with these figures,' Detlef Wittig, Executive Vice President, Group Sales and Marketing, commented in Wolfsburg on Monday. With conditions remaining difficult, the Volkswagen brand even sold more vehicles in April. Compared with the same month in 2008, the core brand delivered 339,500 passenger cars (335,100), a rise of 1.3 percent. For the first four months of 2009, the Group therefore reported deliveries of 1.93 (2.14; −9.6 percent) million vehicles, and the Volkswagen brand reported deliveries of 1.22 (1.26; −3.2 percent) million vehicles. The world market, however, fell by 21 percent during the same period. The Audi, SEAT and Škoda brands also fared better than the market. 'In contrast to the temporary special situation on the German market and the positive trend on the Chinese market, global automotive markets remain very weak and there is no recovery in sight. Nevertheless, we still expect to perform better than the overall market,' Wittig said.... As in previous months, the Group again fared better than its competitors in the Europe sales region in April. While overall sales on the passenger car market in Europe fell by 21.5 percent, the Group reported a significantly smaller drop of 9.1 percent to 294,900 (324,400) units. The Volkswagen brand performed even better with 141,200 (144,600; −2.3 percent) vehicles sold. Developments in France were particularly pleasing. In a market down 7 percent, the Group delivered 23,800 (22,400) vehicles, representing a rise of 6.5 percent, while the Volkswagen brand grew deliveries by 16.0 percent to 13,200 (11,400) units. This confirms the positive trend of previous months and the improved situation in particular with regard to availability of Polo and Golf models.

I know that you may not believe all that you read in newspapers, but an article in the *Times* seems relevant:[3]

The car industry, which has received £2.3 billion in state-backed loan guarantees, has called for the government to introduce a car-scrapping scheme to stimulate demand for new vehicles. Under the scheme, which is understood to have the backing of Lord Mandelson the business secretary, motorists who scrapped a car nine or more years old would be handed a voucher worth about £2,300 towards the cost of a new vehicle. Supporters of the scheme point to Germany, where the introduction of scrapping resulted in a 21% rise in new car sales in February, compared with a 21.9% fall in the UK. A business department spokesman said "We are assessing the cost to taxpayers and the benefits to the scheme."

I found this concept fascinating. There are several thoughts that sprang to mind. Why are there so many different pictures of old airfields, dockyards, and car parks full of new and unsold cars? Why make anything for which you do not have a customer? The other side of the coin is that there are thousands of backstreet vehicle repair garages who are relying on fixing cars nine or more years old to make a living. So, on all sorts of levels, why was this scheme considered?

On the one hand, the introduction of a vehicle scrappage scheme by the British government is undergoing some churn; on the other hand, in Germany it very much looks as if the scrappage scheme has not only stopped the decline but also made a positive difference to their market share. The British scrappage scheme came into effect in mid-May 2009 amid growing concerns from Ford, Honda, and Vauxhall that the scheme was unworkable following a last-minute document from Revenue and Customs, a government agency.

In brief, the British should have provided eligible motorists who owned vehicles older than ten years a subsidy of £2,000 off the list price of a new vehicle. Half of the subsidy came from the government, and half from the motor manufacturers. The British government pledged £300 million for the scheme which was designed to last up to nine months or until the funding was consumed. The last-minute glitch revolved around issues connected with value-added tax (VAT). Under normal circumstances, VAT is paid by the buyer, and the document from Revenue and Customs has inadvertently clouded the issue. Time will tell how quickly the issues are resolved. The longer the delay, the more difficult motor traders and suppliers will find high street trading conditions.

Initial Payment	Amount	Final Cost	Final Amount
First payment	£125	Final repayment (including £165 option-to-purchase fee)	£2,888.50
Additional customer deposit	£0	On-the-road price	£6,795
Government scrappage allowance	£1,000	Scrappage on-the-road price	£4,795
Suzuki GB scrappage allowance	£1,000	Total amount payable: £6,128.50	£6,128.50
Amount of credit	£4,795	APR (typical)	11.8%

FIGURE 5.2
A scrappage financing scheme.

In fact, the British scheme was not identical to schemes elsewhere. In each case, the British government funded £1,000 for the scrappage scheme and required a further £1,000 by the motor manufacturer. In effect, the motor manufacturers "lost out" to the tune of £1,000. I kept one of the new car advertisements on which is printed the following (and see Figure 5.2): "Offer applies to vehicles privately registered between April 1st 2009 and June 30th 2009. Scrapped vehicle must have been registered to the purchaser for at least 12 months."

The issue here is the cost of the loan. At the time scrappage deals were in the marketplace, a saver could expect between 0.5 and 1.5 percent interest from any savings. An interest charge of 11.8 percent to borrow the money for the new car under the scrappage scheme is far higher than the more typical rate offered in the car showrooms (typically, 3.9 percent). One industry commentator suggested that drivers would be better off ignoring the £2,000 grant and obtaining a more competitive deal by selling their own car privately. Not all scrappage APR interest charges were over 11 percent, however: for example, the Ford Kuga deal was 7.9 percent. The Society of Motor Manufacturers and Traders said,[4]

[U]nlike most of European scrappage schemes, which are entirely funded by government, the UK scheme demands an industry contribution of £1,000 to match the Government's own input. In some cases, where the manufacturer profit margins are low, they are not able to offer additional incentives which may still be available on non-scrappage models and this may be reflected in the finance arrangements.

The German motor car buyers, however, have not experienced as much churn as their British counterparts. The initial investment from the German government into their scrappage scheme was €1.5 billion. Since this was announced, a further €3.5 billion was pledged by the German government, bringing their total to €5 billion, which is £4.5 billion at today's exchange rate. More than 1 million cars have been registered in the German scrappage scheme so far, which is estimated to affect over 600,000 motorists. So while March 2009 UK sales fell by 28 percent compared with March 2008, the German figures showed a 40 percent *increase* when comparing like months. Both schemes have now ended. The German scheme was much more effective than the British.

Some critics of the British scrappage scheme have made the following observations:

- Some manufacturers increased the list price of their cars by more than the scrappage allowance before the scheme started.
- Only 14 percent of cars bought under the scheme were manufactured in Britain.
- Twenty percent of the cars manufactured in Britain were exported overseas.
- Perfectly good cars were scrapped, raising environmental concerns as old well-maintained cars create more waste than keeping the old cars roadworthy.
- According to some experts commissioned by a consumer television program, some scrapped vehicles might have gone on to provide good service for at least the next five years.
- Some older cars have been scrapped that are regarded as "classic" cars, which has affected the overall vehicle heritage.

Furthermore, secondhand cars are now more expensive than they were before the scrappage scheme was introduced, as there are now fewer of them in the secondhand car market.

The various scrappage schemes will provide some much-needed breathing space to the motor manufacturers and their suppliers. The question for us here is what will these companies do with that time? If all that they do is breathe a collective sigh of relief and carry on manufacturing operations as they did prior to the subprime crisis, then ultimately they will have gained nothing. Arguably, what they need to do with the time that scrappage schemes provide is to restructure their waste reduction activities so that as the world financial position increases and we come out of this global recession, those who have better implemented lean thinking concepts will be better placed to increase their market share relative to those of other manufacturers.

In other words, this recession is a business opportunity too great to be missed. I would not suggest that the misery and suffering felt by countless millions of people should be taken lightly. I am merely stating that there is a waste reduction opportunity and companies that redeploy or encourage their employees to devote more of their collective energies into waste reduction activities will ultimately have potentially longer careers in their chosen workplaces because the companies that they work for will be better able to grow in the face of the competition. In the first few months, implementing lean thinking will result in work overload, and should help to increase staff morale in these uncertain times.

Of course, not all governments have taken the route of introducing scrappage schemes. The Japanese government is reported to be considering a stimulus package for its economy that will tie subsidies to more environmentally green alternatives. Should this scheme be approved and introduced, there would be incentives not only to exchange vehicles but also to include energy-efficient white goods. It's difficult to confirm the figures until the Japanese government publishes its proposed scheme; ¥15.4 trillion (£100 billion) has been mentioned.

The Japanese Government introduced a scheme for vehicles of 13 years or older. The scheme also covered cars imported into Japan.

5.1 BAILOUT CONDITIONS

No attempt to document here all of the sell-offs, mergers, and bankruptcies in the motor manufacturing businesses that have been caused by the credit crunch will add value. It may be another year before all of the churn has ceased and whatever passes for "normal" returns to the sector. However, it is worth recording here some of the issues that *will* change how lean thinking is perceived in this business sector.

Let's just have a quick look at one specific meeting that took place between the CEOs of Ford, GM, and Chrysler with Congressman Brad

Sherman—at least how it was reported in the British press.[5] The story goes as follows:

> 'I'm going to ask the three executives here to raise their hand if they flew here (to Washington) commercial' said Congressman Brad Sherman. No one stirred at the table where the chiefs of America's big three carmakers had gathered to beg for another $25 billion (£17 billion) bailout for their stricken firms.
>
> 'Second, I'm going to ask you to raise your hand if you're planning to sell your jet and fly back commercial' continued Mr. Sherman as expensively upholstered buttocks shifted uncomfortably on seats. 'Let the record show that no hands went up.'
>
> This scene on Capitol Hill has become one of the defining images of the recession and without doubt contributed to Congress's reluctance to agree a rescue package for the carmakers this week.

The news article continued,

> 'There's a delicious irony in seeing private luxury jets flying into Washington, DC and people coming off them with tin cups in their hands' said another, Congressman Gary Akerman. 'It's almost like seeing a guy show up at the soup kitchen in high-hat and tuxedo' he added.

The story unfolded further at the end of March 2009, with GM Chairman and CEO Rick Wagoner becoming a high-profile casualty of government intervention following thirty-two years with GM. He was replaced as CEO by Fritz Henderson, the company's vice chairman.

As other U.S. government representatives and U.S. motor industry executives met to thrash out further government funds, GM posted an amended viability plan on its website. It contained the following statement:[6]

> The Viability Plan announced today builds on the February 17 Viability Plan submitted to the U.S. Treasury. The revised Plan accelerates the timeline for a number of important actions and makes deeper cuts in several key areas of GM's operations, with the objective to make us a leaner, faster, and more customer-focused organisation going forward....
>
> Significant changes include:

- a focus on four core brands in the U.S.—Chevrolet, Cadillac, Buick and GMC—with fewer nameplates and a more competitive level of marketing support per brand
- a more aggressive restructuring of GM's U.S. dealer organisation to better focus dealer resources for improved sales and customer service
- improved U.S. capacity utilization through accelerated idling and closures of powertrain, stamping, and assembly plants
- lower structural costs, which GM North America (GMNA) projects will enable it to breakeven (on an adjusted EBIT (earnings before interest and taxes) basis) at a U.S. total industry volume of approximately 10 million vehicles, based on the pricing and share assumptions in the plan. This rate is substantially below the 15 to 17 million annual vehicle sales rates recorded from 1995 through 2007

... 'We are taking tough but necessary actions that are critical to GM's long-term viability,' said Fritz Henderson, GM president and CEO. 'Our responsibility is clear—to secure GM's future—and we intend to succeed. At the same time, we also understand the impact these actions will have on our employees, dealers, unions, suppliers, shareholders, bondholders, and communities, and we will do whatever we can to mitigate the effects on the extended GM team.'

The bailout plans as stated helped to secure more funding for GM with Ford and Chrysler also issuing statements regarding their own short- and medium-term business plans. Meanwhile, mergers, sell-offs, and bankruptcy talks in the rest of the industry continued.

At roughly the same time, there were a few comments by President Barack Obama released on 20 May 2009 during his first quarterly meeting of the President's Economic Recovery Advisory Board, which took place in the Roosevelt Room:[7]

But obviously one of the things that I've been concerned about since I took office is looking beyond the immediate crisis in front of us to find out what is a sustainable economic model post-bubble and bust. How do we create sound fundamentals on issues like education, on health care, and the topic that we're going to discuss today, energy, as well as all the innovation that's required around these various areas, so that moving forward we don't find ourselves in an unsustainable economic model?

And we have seen this week some fairly extraordinary steps being taken around energy, which are promising. Yesterday I stood out in the Rose Garden and announced that the automakers, the unions, state and local

	Miles per U.S. Gallon	
	City Driving	**Highway Driving**
Fusion Hybrid	41	36
Yukon Hybrid	21	22
Ford Mustang	18	26
Chrysler 2.6L DOHC V6	18	26
Chrysler 2.7L DOHC V6	18	26
Ford Taurus X	17	24
Cadillac STS	13–17	19–26
Cadillac SRX Crossover	13–15	20–23
Chrysler 6.1L SRT	13	19

FIGURE 5.3
Some 2009 fuel efficiency data.

officials, as well as the federal government, were coming up with a uniform national fuel efficiency standard that will provide certainty to the automakers and take a real bite out of our level of oil dependence and over time reduce our dependency on foreign oil.

His comments concerning the motor industry laid out some dramatic new standards on emissions and fuel efficiency targets for future generations of motor vehicles. These targets will compel the most iconic vehicle brands in U.S. history to make fundamental engineering changes to their automotive design. Arguably, this new policy will also force a move to hybrid cars and light trucks. They will need to be 40 percent cleaner and more fuel efficient. The new rules come into effect in 2012 and stipulate that by 2016, most passenger cars must reach 39 miles per U.S. gallon (47 miles per imperial gallon), with equivalent increases in fuel efficiency for light trucks.

Figure 5.3 shows some current fuel efficiency data (2009) for a selection of the 2009 GM, Ford, and Chrysler vehicles. It illustrates the challenges facing automotive design as designers look for technology to bridge the efficiency gap.

These data were obtained from the relevant motor manufacturers' websites and are official figures, expressed in miles per U.S. gallon. Just to place these data in context (Figure 5.4) shows the data for some of the well-known hybrid vehicles manufactured by non-U.S. companies, and Figure 5.5 shows some of the most fuel-efficient vehicles worldwide. The Citroën C1 MPG

	Miles per U.S. Gallon	
	City Driving	Highway Driving
Toyota Prius	48	45
Honda Civic Hybrid	40	45
Nissan Altima Hybrid	35	33

FIGURE 5.4
Hybrid mileage.

delivers 68.9 miles per gallon (mpg), some 3.5 times that of the Chrysler 6.1L SRT 19's mpg (in the same units). There's a big difference!

Not only are the American motor manufacturers producing vehicles with very low miles per gallon compared with the rest of the world, but also they are not responsive—at least in the main—to improvements in emissions. The two usually go together, but there may be some exceptions.

Before we look at some of the market disruptors that will also impact the ability of the motor industry to improve its collective lot, let's briefly consider how this anticipated turnaround is going to be achieved. The usual soft targets are to restructure using the short-term measures of workforce reductions, plant closures, and selloffs. Provided a buyer can be found, selling off parts of the business can show a profit in the short term, but you can't shrink to success. Another option for the motor industry is economies of scale, concentrating on the products that sell the most, and axing production of those less profitable vehicles.

Some or all of the above measures will produce some or all of the desired short-term goals. Unfortunately, they will not provide the manufacturing infrastructure to compete against the likes of Toyota, who has always had much longer term strategies. While Toyota has also reduced production volumes to cope with demand, it has been working to lean thinking

	Miles per U.S. Gallon	
	City Driving	Highway Driving
Citroën C1 MPG	44.2	68.9
Vauxhall Corsa-MY2006 MPG	34.9	65.1
Fiat New-Panda MPG	34.9	55.9

FIGURE 5.5
Some 2009 European fuel efficiency data.

principles and is ideally placed to pick up the pace as we come out of the recession and greater stability returns to the marketplace.

If I were in charge of a motor-manufacturing industry, I would factor lean thinking into my short- and long-term plans, working more aggressively now than ever before to reduce waste in all its forms, as well as all of the usual scenarios for turning companies around that are mentioned in this chapter. Consider this: a restructured and smaller company that still has all of its old habits will now be competing against a bigger company (Toyota), the management team of which knows all about waste reduction and long-term success. In fact, I will go further than that and suggest that we should restructure some of the lean thinking concepts so that we are not just following Toyota by implementing the TPS principles, but enhance lean thinking so that we have the opportunity to leapfrog their concepts. At least that way we will have a fighting chance of continued growth. I'll pick this point up further in Chapter 10.

We've looked at the American motor industry in some detail; however, all of the scenarios mentioned are also evident in many business sectors including the finance institutions. Before we wrap up this chapter, let's have a look at further market disruptors to the motor industry, because the rest of the world will not stand still and let the large companies fight among themselves for market share.

5.2 NEW TECHNOLOGY

Let's have a quick look at Yutaka Matsumoto and his team of engineers. Matsumoto works for Toyota, and it was his team that developed the Prius, arguably the car to beat in terms of green credentials. Over a million Prius cars have sold since its launch in 1997. Matsumoto is currently driving an electric car that may not be seen in showrooms for a number of years as he is driving the first plug-in hybrid car. Matsumoto is quoted as saying, "If I use this car for commuting in Tokyo, I can travel at up to 60 mph but there are no emissions whatsoever."[8]

Of course, there are many technical problems to overcome before the technology used in this car will be a commercial success, not least of which is the proposed length of car journeys before recharging. The predicted maximum journey length is fifteen miles or so between charges with current battery technology. As we all know, this is a short journey in American terms but is greater than most commuter journeys in Europe.

The work of Gerbrand Ceder, a professor of materials science at the Massachusetts Institute of Technology (MIT), on the other hand, may help the Toyota plug-in car project. Ceder summarizes his research group's interest on his MIT website:[9]

> Professor Ceder's group specializes in designing and understanding advanced materials by means of computational modelling and experimental research. By combining theoretical and experimental efforts in one group, the effectiveness of both is enhanced. First principles computations, whereby the properties of materials are predicted from basic physics, has become one of the most powerful tools in Materials Research and Design. This group develops these tools and applies them to technologically relevant problems, often in collaboration with key industrial or government partners. Materials phenomena include: phase stability and cohesion in solids, diffusion, interaction of matter with radiation, and phase transformation. Applications have included: high temperature superconductors, electrodes for rechargeable batteries, and high temperature alloys. The environment is highly multi-disciplinary, containing students with a range of backgrounds making use of cutting edge techniques from such fields as materials science, engineering, chemistry, physics, computer science, and mathematics.

At first glance, you might think that this research group has little to do with the real world. However, Ceder has made some comments in the press that help the layperson to understand his research: "if you can charge your phone in 30 seconds, that becomes a life changer," and "it could change the way we think about technology like this: you would literally be able to charge up while you stand and wait."

The technology has acquired the nickname *beltway battery* because it is a bypass system to let the lithium ions that carry charge to enter and leave the battery more quickly. Let's not worry about the details, but merely reflect that technology to solve some of the hybrid car issues is being conducted, and not just at MIT either. There is a further article concerning Professor Ceder and Byoungwoo Kang's research in *Chemistry World*, the member magazine of the Royal Society of Chemistry (Ceder and Kang, 2009, p. 24). This article also cites Peter Bruce, an electrochemist at the University of St. Andrews. Bruce makes the comment that where other research groups have tried and failed in this technology, the Ceder–Kang technology shows promise. Although this technology is some years off production readiness in commercial vehicles, exciting research is taking place in all sorts of locations.

As you might imagine, there are some parts of the world where electric cars are heralded with open arms. For example, Hawaii is set to become the first U.S. state to create a transport infrastructure that will allow cars to run almost entirely on electricity. The plan involves building up to 100,000 charging stations by 2012 and importing electric vehicles manufactured by Nissan and Renault. The infrastructure for this venture is funded mainly from an organization called Better Place.

Better Place is run by Shai Agassi, the world's leading electric vehicle services provider. They aid in building and operating the infrastructure and systems to optimize energy access and work with relevant players to create an integrated solution.[10] The electricity for the whole scheme is expected to come from renewable sources such as wind power. Ultimately, the potential is for electric vehicles to cost less than equivalent petrol cars due to fewer moving parts. The Better Place model is interesting and has attracted much interest with potential schemes in the San Francisco Bay area. Israel, Denmark, and Australia are also emerging as potential future partners.

More recent technology has been jointly developed between Better Place and Nissan. Figure 5.6 shows the concept, which was demonstrated in

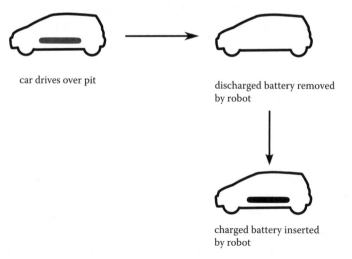

car drives over pit

discharged battery removed
by robot

charged battery inserted
by robot

FIGURE 5.6
Robotic battery replacement.

2009 in Yokohama. The demonstration showed the robotic replacement of the discharged battery undertaken in a car park and not a filling station. For this system to be universal, there would need to be industry-wide agreement for all batteries to have the same dimensions and charge characteristics. Otherwise, there would be a plethora of battery dimensions and electrical characteristics which would not benefit the consumer or our carbon footprint. The exciting feature of the replacement is that it is achieved in under a minute, which is perhaps the normal length of time that we wait at a petrol filling station.

The idea is that the batteries are rented somewhat like propane canisters for welding, barbecues, and so on. It's a great idea with much to recommend it. Should this technology succeed, it will be interesting to watch the process of battery removal undertaken on a vehicle that has been driven in cold conditions where the roads have been "salted." The underside of these vehicles will have many rust points which may hinder battery removal. Under these conditions, it may be the case that the battery-retaining bolts are difficult to remove or indeed shear off. Agassi will doubtless overcome this problem like he has the other obstacles.

In a supportive move, Sir Terry Leahy, Tesco's chief executive, has announced that two London stores will pilot electric car–recharging facilities. The idea here is that your car will be recharged while you shop in the supermarket. In a speech at the London School of Economics, Leahy commented that "businesses have to show how the consumer can make a difference … the danger is people are so taxed already that every time green is mentioned it is with a new tax that just switches people off the whole subject."[11]

Tesco is also intending to build the world's first zero-carbon store in Ramsay, Cambridgeshire.

Mayor of London Boris Johnson announced in April 2009 his plans to install 25,000 charging points by 2015 to service 100,000 electric vehicles. Additionally, the Optare Bus Company is manufacturing electric buses that emit no pollution and are cheaper to run than their diesel equivalents. Battery life is a problem in that the buses can travel for only 60 miles before they need to stop for a 15-minute recharge. Were these buses to be introduced for fee-paying passengers, the relevant timetables will need to be altered to allow for recharging.

There is one potential safety hazard in that the buses are almost inaudible, except for some minor tire noises. Optare is considering introducing a speaker system so that pedestrians are aware that there is a moving vehicle nearby. A recent newspaper article reported news of a company that was hoping to sell suitable sound systems that would broadcast noises from under the electric car bonnet. At one stage, they were proposing the engine sounds of top-end, high-performance cars such as Porsches and Bugattis!

Small electric vans have also taken to the streets in a variety of cities, with a leading British supermarket and two parcel carriers taking delivery of some of the first commercially produced Modec vans. Modec has costed their vans at 15 British pence (15p) per mile compared with 37p per mile for conventional vans. Starting from a base of 100 vans in 2008, company founder Lord Borwick expects numbers to increase to 350 in 2009 and 1,200 in 2010. Lord Borwick also tried to manufacture an electric taxi but found that many London taxi drivers travel many miles into London before they even start to pick up fares. At the moment, this market sector cannot convert to electric power until battery technology ensures greater miles between charging. Incidentally, Modec operates to lean thinking techniques, as their website demonstrates:[12]

> Heralded as the future of commercial transport, the British manufacturer was founded in 2004 by a team of automotive engineering experts and its vehicles were launched to the transport industry in March 2007. The Coventry based company is leading the field by taking automotive technology to the next level, using the very latest in battery technology and LEAN manufacturing techniques.
>
> Modec is expanding rapidly. International dealers are now being established in France, The Netherlands, Spain, Germany and Ireland. This growth is set to not only transform the commercial vehicle sector, but the whole automotive industry.

Introducing electric vehicles in any numbers will take a long time, as attractive scrappage deals are made available by central governments, and the recession changes the buying public's ability to invest in new cars of whatever type. There will be casualties. Indeed, a British concern, the Nice Car Company,[13] which was only one of two British distributors of electric cars, went bankrupt at the end of 2008. The British government is intending to provide grants for the purchase of electric cars, which will be approximately the same as the current scrappage deal for fossil fuel cars; the scheme comes into effect in 2011.

Also, there are many technical problems to overcome. A personal concern of mine would be that we are using electricity to clean up the motor vehicle emissions. We need to ensure that we are also cleaning up power station generation, which is well outside of the scope of a book concerning lean thinking.

Just this brief survey of electric vehicles demonstrates that the market is moving at a fast rate and the pace will quicken further with time. More companies will consider electric vehicles, and yet others will consider alternative fuels. For example, the spring 2009 edition of the in-house publication *Advanced Driving*, which is written and published by the Institute of Advanced Motorists, contains an article concerning the Honda FCX Clarity. The article starts,

> [I]t has been hailed as Car of the Century; the beginning of an automotive revolution and the first nail in the coffin of the carbon economy. And as the world's first production hydrogen-fuel-cell vehicle. Honda's FCX Clarity could well be the most important car since the Model T.

While the article in *Advanced Driving* might be difficult to obtain, searching the web for "Honda FCX Clarity" produced over 250,000 hits. There are plenty of other reviews saying much the same thing.

There are 200 drivers in California who are taking part in test drives, including the actress Jamie Lee Curtis. Refilling the hydrogen tank is now easier than it used to be as there are currently five hydrogen filling stations in the greater Los Angeles–Orange County area, with Honda expecting 16 by 2010. The latest information (early 2011) is that there will be 24 stations in Los Angeles, 8 in Orange County, 8 in San Diego, and 10 elsewhere in California by 2015. The hydrogen in conjunction with atmospheric oxygen produces electricity, which is stored in fuel cells in the rear of the car.

As every school student will know, burning hydrogen with atmospheric oxygen also produces water, so the car is indeed as "green" as can be in terms of emissions. The seats are made from a textile derived from corn, and so the green concepts also apply to as many of the other components as is possible using today's technology.

The only issue here is the manufacturing process used for generating hydrogen. A process needs to be found that is green; otherwise, the problem of greenhouse gases will just shift to a different part of the overall energy generation cycle. Whatever the eventual technology, it also needs to be cheap if there are to be sufficiently large numbers of hydrogen-powered vehicles.

There is a recent report in *Chemistry World* (June 2009) of some chemists at the University of Amsterdam who can produce hydrogen from methane combustion during power generation. The new catalysts used in this process are based on ceria (CeO_2). When doped with such metals as platinum, ceria helps to break down methane into hydrogen. Cheaper metals including nickel have also been tried and proved to be effective. Of course, time will tell if these new catalysts will be of long-term benefit. Additionally, the power used here is still derived from power stations; nevertheless, new technology may produce a breakthrough that in the longer term helps to bring the price of hydrogen down without impacting the carbon footprint.

So there is competition for the Detroit motor industry from many areas of the world of electric vehicles, but what of the moves to make petrol-driven cars more accessible to some parts of the world? For example, the British government established the Carbon Trust as an independent company in 2001 with the aim of accelerating the move to a low-carbon economy. They state on their website,[14]

> We cut carbon emissions now:
> - by providing business and the public sector with expert advice, finance and accreditation
> - by stimulating demand for low carbon products and services
> We cut future carbon emissions:
> - by developing new low carbon technologies through project funding and management, investment and collaboration
> - by identifying market barriers and practical ways to overcome them

The Carbon Trust established the Carbon Trust Standard Company in June 2008, with the Carbon Trust Standard certification created to provide a much-needed clear and robust definition of good practice and an independent endorsement of an organization's achievements in carbon reduction: "We work with leading organisations to certify their performance in measuring, managing and reducing their greenhouse gas emissions."[15]

The increased use of biofuels mixed with standard fuels may not necessarily help with carbon emissions. While it is the case that biofuels absorb oxygen as the crops of oilseed rape, palm, and soya are being grown, some fields where these crops have been planted were created by destroying forests. Additionally, older vehicles are unable to cope with higher levels of biofuels mixed with petrol. Some of the issues relate to rubber seals that perish with the increased levels of ethanol (the active ingredient of biofuels), and the corrosion of some parts. In either event, there will need to be considerable work undertaken by the Detroit carmakers before significant levels of biofuels are used in the larger vehicles.

Perhaps one of the most interesting recent events is that of Tata,[16] which has launched extremely cheap cars for the Indian market. The Nano has been extremely well received in India even though there are only basic features for the three models announced. These vehicles have small engines and much-reduced features relative to vehicles produced in the West. Stripping away unnecessary weight and features increases fuel economy and directionally helps carbon emissions. It certainly is an alternative approach!

So how can one sum up some of these technologies that will hamper the Detroit motor companies as they recover from the credit crunch of 2008–2009? It will be a challenge just to stay in business. Indeed, at the time of writing this book (late 2009), GM made the biggest industrial bankruptcy filing. The White House commented that the carmaker might use Section 363 of the U.S. Bankruptcy Code to sell its assets to a government-funded company that would emerge as the new GM.[17] In addition, GM is close to a deal to sell off its Hummer brand and recently pulled out of a deal to sell its European interest, known in Europe as Opel and Vauxhall. The GM management chose not to sell its European manufacturing base as the scrappage scheme had created a much-needed boost to their factories. Whatever the final outcome, the old GM will not be able to recover its once-dominant position easily, if at all. Chrysler filed for bankruptcy protection in April 2009, so their long-term recovery will also be interesting.

So here we have two of the big three Detroit companies in need of restructuring facing competition from their traditional rivals such as Toyota and the like, and the increasingly important greener technologies. As the Detroit companies work through these issues, what is the likelihood that they will embed lean thinking techniques into their leaner and fitter businesses? Time will tell, but it does not look promising.

NOTES

1. http://money.cnn.com/magazines/fortune/fortune500/2009/.
2. https://www.volkswagen-media-services.com/medias_publish/ms/content/de.standard.gid-oeffentlichkeit.html.
3. http://www.timesonline.co.uk/tol/sitesearch.do?query=scrappage&turnOffGoogleAds=false&submitStatus=searchFormSubmitted&mode=simple&sectionId=674.
4. http://www.smmt.co.uk/home.cfm.
5. http://business.timesonline.co.uk/tol/business/industry_sectors/engineering/article5209281.ece.
6. http://preprodha.ecomm.gm.com:8221/us/gm/en/news/govt/docs/plan.pdf.
7. http://www.whitehouse.gov/the_press_office/ObamaAnnouncesEconomicAdvisoryBoard/.
8. http://www.timesonline.co.uk/tol/news/environment/article6168857.ece.
9. http://dmse.mit.edu/faculty/faculty/gerd/.
10. http://www.timesonline.co.uk/tol/news/environment/article6349906.ece.
11. http://www.businessgreen.com/business-green/news/2243260/tesco-trial-electric-car.
12. http://www.modeczev.com.
13. http://www.nicecarcompany.co.uk.
14. http://www.carbontrust.co.uk/Pages/Default.aspx?gclid=CPCOs9rF1p8CFYSX2AodWVr8bw.
15. Ibid.
16. http://www.tatamotors.com.
17. http://www.whitehouse.gov/the_press_office/Fact-Sheet-on-Obama-Administration-Auto-Restructuring-Initiative-for-General-Motors.

REFERENCE

Ceder, G. and B. Kang. 2009. *Chemistry World*, April, p. 24.

6

How the Toyota Production System and Lean Thinking Are Currently Implemented

The Six Sigma methodology of waste reduction has the initial target of 3.4 defects per million opportunities. Lean thinking, on the other hand, is concerned with producing zero waste by the continual erosion of the seven causes of waste as outlined by Taiichi Ohno. Yet there is a connection between the two quality systems in that Six Sigma change agents have always been known by the "belt" terminology of *greenbelt*, *blackbelt*, or *master blackbelt*. The belt training now incorporates most if not all of the lean thinking methodology. So there can be, and indeed is, some confusion from time to time. Let's have a look at the origins of the terminology.

6.1 "BELTS"

Over the last twenty years or so, many companies have tried to implement Six Sigma methodology. A complete implementation would have the "sigma level" increasing by one unit every three years provided that the rate of improvement was 54 percent each year, year on year, during the implementation. There were not many companies that established that rate of defect reduction, certainly not across a whole company.

What's the point of mentioning the early Six Sigma implementations? Well, the specialists who Mikel Harry trained were known then as *blackbelts*. Harry coined the term in the mid-1980s while he was consulting for the circuit board manufacturer Unisys Corporation. At that time, the term was used to refer to project leaders trained in statistical problem solving. As one might imagine, the roots of the name lie in the similarities between martial arts and Harry's breakthrough strategy.

Harry described the necessary skills needed to be a blackbelt as people who would "stimulate management thinking by posing new ways of doing things, challenge conventional wisdom by demonstrating successful applications of new methodologies, seek out and pilot new tools, create innovative strategies and train others to follow their footsteps" (Harry, 2006, p. 203).

He then made comments about the character of the people who should be recruited to be blackbelts:

> [T]hey have to be patient, persuasive, imaginative and creative. They need the respect of front-line employees, those in supervisory positions, and middle and senior management if they are to manage risk, set direction and pave the way to breakthrough profitability. Most important of all they translate intention into reality and help to sustain it. They clarify an organisation's purpose and funnel that clarity into specific blackbelt projects. (2006, p. 203)

Sound familiar?

Harry then went on to define a *greenbelt* as someone who works part-time in his or her specific areas and part-time on a given improvement project. He also defined a *master blackbelt* as an individual who was responsible for transferring Six Sigma expertise to blackbelts. So the master blackbelt is an individual with many blackbelt projects to his or her name and many years of experience in implementing Six Sigma concepts.

The term *blackbelt* caught on to such an extent that it became the industry standard terminology for a process improvement specialist, and organizations set themselves up training individuals in Six Sigma techniques. Large organizations also set up their own in-house blackbelt academies. Eventually, each of the training organizations settled on a common format and to some extent reached common ground concerning the training course content so that the term *blackbelt* became synonymous with specialist-specific industry-wide training and experience, so much so that some companies began recruiting blackbelts trained by any one of a number of organizations.

Of course, this didn't last. Slowly over the years, the training course content was adapted and modified so that now the training courses tend to be four or five weeks in length and cover lean thinking, TPS concepts, and, incidentally, fewer concepts from the original Six Sigma methodology. In other words, the term *blackbelt* has been hijacked by the lean thinking movement. This is not a problem but does explain why some communities refer to *lean* as *lean Six Sigma*.

There is another belt classification that needs to be discussed: the management blackbelt. This is the training that I undertook. In my day (the early 2000s), training was designed to bring managers up to a level of lean thinking understanding so that they formed a team able to follow lean principles in order to achieve the desired objective. I'll cover the blackbelt training in more detail in Section 6.2.

6.2 BLACKBELT TRAINING

What is all this talk of "certified blackbelt training" or "certified greenbelt training"? Which body worldwide is responsible for certifying the trainers? I am really not out to question any firm that offers certified training. What I would like to understand is the process leading up to certification within the many training organizations. In other words, is there a worldwide body somewhere that controls the course content of blackbelt training and ensures consistency in terms of both breadth and depth of training? If so, where is this company or institute? Actually, I am still looking for one!

The serious point to make is about progress. As we shall see in Chapters 7–9, there are all sorts of new concepts in business psychology and engineering that are technologies or techniques worthy of inclusion in the "belt" training. What's the mechanism for their inclusion? It's a theme that I will develop in Chapter 10.

Let's have a look at the course content from an in-house blackbelt course that ran just a few years ago. This will not be a detailed account since training courses will vary from year to year in their examples and also from company to company. The idea here is to get a feeling for the subjects that are covered in a typical blackbelt course.

For those who have not been involved, in-house blackbelt training is often one week of theory, followed by three weeks of the project, followed by a further week of training and then another three weeks of project, and so on. The course therefore lasts for five months, most of which involves the course participant working on his or her own blackbelt projects. Multinational companies running their own in-house training often run these courses in different locations so that travel is kept to reasonable levels. There is nevertheless a huge undertaking not only by the course participant but also by the company.

Additionally, blackbelt training projects are required to have clear deliverable goals the savings from which can be measured and translated into real monetary savings. The nature of the projects is such that they may take several months after the last of the formal course training weeks before project completion. Start to finish could therefore take the best part of a year. It seems and indeed is a long time. However, those projects that I have observed had financial returns in six figures (doesn't matter whether it is measured in pounds or dollars). Such a large return pays for the course in the first year, leaving trained blackbelts to undertake future projects with few further training costs.

Let's have a look at the course material (see Figures 6.1, 6.2, 6.3, 6.4, and 6.5). Exercises include catapult (data generation and analysis) and role play.

"Putting It All Together" is a summary afternoon toward the end of week 5 during which the course leaders help the participants organize their thinking and point them in the right direction. This is a necessary feature of any training program. As mentioned earlier, this training course takes months. A lot can happen to an individual and organization during

Communication style assessment	Thought process map
Understanding communication preferences	The stages of team evolution
Team roles	Communication styles
Understanding clients	Self-improvement
Listening and interpreting	Successful blackbelt interventions
Interpersonal skills	Effective decision making
Selling Six Sigma and bidding for resources	Meeting skills
Best practice for consultants	Practical application
Process skills	

FIGURE 6.1
Week 1.

Employee hiring process	Cycle time
Baseline methodology	Six Sigma measures – defect and yield measures
Voice of the customer	Management by fact
Collecting and summarizing data	Measure your process
Project contract reviews	The 6-sigma paradigm shift
Cost of quality	Project management overview
Process mapping	

FIGURE 6.2
Week 2.

Analyze your process overview	Failure modes and effects analysis (FMEA)
Data-based decisions	Lean thinking
Systems thinking	Z-charts (a combination chart)
Control charts—concepts	Process capability
Control charts—interpretation	Measurement system evaluation
Control charts—construction	

FIGURE 6.3
Week 3.

Improve your process	Design of experiments
Inferential statistics	Screening analysis and interpretation
Modeling	Optimization
Reliability	

FIGURE 6.4
Week 4.

Control overview	Pugh's concept selection (decision matrix method)
Tolerance analysis	Analysis methods (FAB-PV)
Robust design introduction, planning, and analysis	Value management
Decision and risk analysis	Scenario planning
Concurrent design	

FIGURE 6.5
Week 5.

that timescale, so this is a vital part of the process. Obviously, each week contains a summary session helping the participants to share experiences prior to sending him or her out into the big world.

The course is broken down into different topics or themes, and can be described in the following way. Week 1 is essentially "soft skills," or those skills necessary to be effective when dealing with self and others. A lot of material is covered in a short timescale. Other courses vary in their content, but most cover similar material. I personally know only twenty or so blackbelts. Most if not all have science or manufacturing backgrounds—many being recruited following a stint in a manufacturing department. They have all benefited from soft skills training.

Clearly, manufacturing experience is relevant to some multinational organizations and not to others. From my own experience in dealing with blackbelts, I know that very few have sociology, psychology, or business training. I might suggest that a week concerning people skills for potential blackbelt practitioners with these backgrounds is insufficient for their needs. On the other hand, the course is long enough and something else would have to give—either the timescale or course content.

Another issue of concern is the personality type of the course participant. An extraverted thinker would absorb this material to a different extent than an introverted feeler. These terms for personality types have their origins in Jungian psychology and will be covered in more detail in Chapter 8. In Belbin team role terms,[1] a company worker may take on board a different perspective compared with the shaper, for example.

Week 2 is essentially an introduction to scientific methodology, trying as far as is possible in one week to help the course participant to grapple with scientific processes of data collection and analysis. A course participant with research experience, particularly the formal training of master's or doctorate degrees, will be well suited to the basic concepts. Listening to the customer is a vital role for any team leader, or indeed team member, no more so than in the blackbelt environment where the basics of listening and reacting to the customer's end goal are vital. Throughout the week, there are exercises helping to consolidate the body of knowledge. Additionally, the participants use the relevant people skills covered in week 1. Not only does this consolidate the week 1 material, but also it helps to establish behaviors.

Week 3 starts the process of bringing all participants up to the same level of understanding in some of the statistics needed to separate common cause from special cause variability. This is important in sorting out one-off occurrences from basic trends in data. In this particular training course, this is the first week during which lean thinking concepts are actively discussed. That's not to say that the concepts of lean thinking are not delivered in the previous weeks—merely that the concepts now have a name and a place.

There is so much material within week 3 that week 4 essentially probes some of the concepts deeper. *Design of experiments* is always a term that causes me some amusement. Having worked for three years for a doctorate and three subsequent years as a postdoctoral demonstrator, I can say without a doubt that all experiments are designed. Unfortunately, some experiments are poorly designed and some are designed with a high probability of success! Many software packages are now available that will help in both design and analysis of an experiment. So this module is needed. Unfortunately, in the real world there may be less opportunity to generate all of the data that may be required for this type of statistically designed experiment to be effective. So in some cases, an element of judgment must temper the experiment.

Week 4 steps back from the coal face of data collection and analysis to look at some overview concepts of decision and risk. It also helps the potential blackbelt to ask relevant questions and temper activities to deliver the best "bang for the buck." It may be common sense to tackle those areas that will most easily return a positive outcome. However, common sense is by no means "common," and some of the training in week 5 helps to "ground" the individuals.

So there we have it: one example of a practitioner blackbelt training course. You may have read the last page or two and asked yourself, "Why spend this amount of paper and ink discussing the various training elements?" The answer can be found in more detail in Chapter 10. However, the brief answer is that once someone is trained, there is no refresher training, nor is there an opportunity to obtain further skills within the "belt" program. Indeed, there are many blackbelt practitioners who received this training and have been active for a number of years on various projects with no further training in lean thinking. Whether this is a good or bad thing will be covered in Chapter 10.

Greenbelt training takes less time—it can be a four-week course spread out over four months, with a project whose expected return for the funding organization is less than that of the blackbelt projects. Additionally, most greenbelts will be asked to undertake their assigned greenbelt duties on a part-time basis. This has benefits since these individuals face the day-to-day issues associated with the "process" whether that is in manufacturing, the finance sector, or wherever. These individuals will know or be themselves operators of their respective processes. They are a vital tool in any organization. We saw in Chapter 2 that a plant manager in Leeds (Mike Harding) met with astonishing success for his factory by employing many greenbelts rather than a small number of blackbelts. So there is a model of success to training and deploying these individuals.

So what does greenbelt training consist of? In the main, courses are often based around the DMAIC concepts mentioned earlier; see Figure 6.6 for more details. This simple and easy-to-remember mnemonic helps the practitioner to remember five simple steps that cover a lot of ground in terms of analysis and improvement. The arrows in Figure 6.6 are drawn to signify that this is a never-ending loop, for there is always improvement to make!

Many of the concepts covered in blackbelt training are mentioned or covered in detail during the greenbelt training. This is not a surprise because the basic concepts apply however they are implemented. Furthermore, it may be the case that the greenbelt practitioner goes on to take a course converting his or her proven skills into a blackbelt qualification. So it is wise from a number of perspectives to design both courses with common weekly goals.

While practitioner blackbelt and greenbelt training is designed for workers who are likely to remain specialist change agents, the training provided for management blackbelts recognizes that course participants have full-time careers in management. Managers are unlikely to switch careers to become full-time change agents, and so the management blackbelt training has different aims and expectations. By way of an example, let's have a look at the training that I undertook a few years ago. The training was full-time for one week followed by a project that in my case had to deliver at least $100,000 worth of improvement. At the time I undertook the training, I was working for a multinational corporation with headquarters in the United States. The blackbelt training program was administered from corporate training that was also based in the United States. The financial goals and deliverables from each of the "belt" projects were therefore all measured in the U.S. dollar.

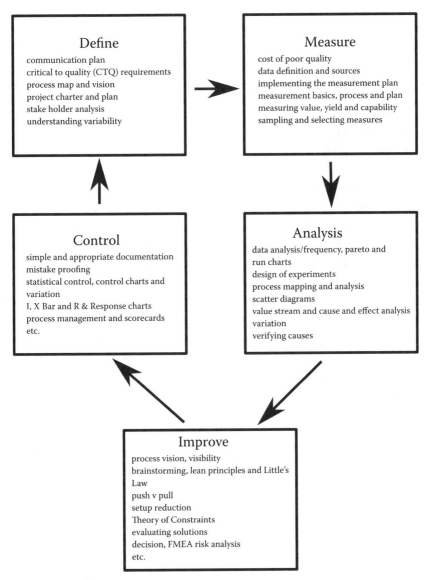

FIGURE 6.6
The DMAIC process.

The course content consisted of the following topics:

- Statistical thinking
- Analyzing managerial data
- Baselining
- Benchmarking

- Management by fact
- Six Sigma fundamentals
- Building winning products
- Lean concepts
- Design of experiments and robust design
- Decision and risk analysis
- Scenario planning
- Managing improvement activities

There were a couple of practical exercises each day to help to consolidate the classroom-based activities. Most of us in the course had been exposed to some of the concepts in previous training, particularly the Six Sigma concepts. Indeed, I was responsible for working with a trainer from the University of Motorola and organizing one- and two-day training courses for Six Sigma in the early 1990s.

As you might imagine, this course could do no more than provide an understanding of the types of training that were being provided to those members of our own departments who we were asking to undergo the practitioner courses. The course also brought our own knowledge up to date and provided us with a common language and shared experience. It also saved the company money!

NOTE

1. http://www.belbin.com.

REFERENCE

Harry, M., and R. Schroeder. 2006. *Six Sigma: The Breakthrough Management Strategy Revolutionizing the World's Top Corporations.* New York: Currency. ISBN-10 0-385-49438-6.

7

Recent Advances in Process Understanding

We've seen earlier that the principle of "stop the process" (*jidoka* in lean thinking terminology) actually derived from the weaving industry and been successfully applied to many industries as part of lean thinking. Taiichi Ohno's adaptation of this concept for use in the Toyota Motor Company was revolutionary. So if Ohno can adapt, modify, and then develop concepts from the spinning and weaving industry, what else is out there in other manufacturing industries that might be capable of being generalized and applied to any process improvement—be that process in the financial, manufacturing, or service industries?

Well, by way of just one example, the British government sponsored an initiative called the Technology and Innovation Futures project (often shortened to Foresight), launched in the early 1990s. In its report called "Energy_1994–99," which is available from the Department of Trade and Industry (DTI) Foresight website,[1] the report comments,

> The Technology Foresight Program is a major initiative which was announced in the 1993 White Paper 'Realising Our Potential'. The Program brings together industry, academia and Government to consider how the UK can best take advantage of opportunities to promote wealth creation and enhance our quality of life. The Program has been driven forward with great energy and enthusiasm by the 15 independent Technology Foresight panels. The Program has reached out to over 10,000 people.

The Foresight panel of interest here is the Energy Panel, which set out its most important areas (see, e.g., Dyer et al., 2008) of interest as follows:

- Reservoir engineering
- Photovoltaic materials and manufacturing
- Gasification
- Low-emission combustion systems
- Gas turbo-machinery
- Batteries
- Process design
- Building services and design

The first round of Foresight was concluded in early 1999, and its summary report[2] listed the following among its inventory of achievements:

Foresight Challenge funding of £17 million for five initiatives, of which the key ones were

- the Institute of Applied Catalysis, the Centre for Protein Technology and the Centre for Process Analytical and Control Technology
- Foresight Link awards, amounting to £1.9 million, notable among which was the Laboratory-on-a-chip project
- improved funding for chemistry research via Engineering and Physical Sciences Research Council (EPSRC)
- establishment of the UK Analytical Partnership to strengthen the UK's analytical science
- increased multi-disciplinary interactions at three major interfaces:
  - chemistry/biology, chemistry/materials, and chemistry/chemical engineering

Numerous associate programs were established, stimulating awareness of the Foresight Program and its importance to business competitiveness. The Foresight key issues for their second round are listed as:

- resource conservation
- social issues
- chemistry at a molecular level

Foresight is now a part of the Government Office for Science, which is headed by Professor John Beddington, the government chief scientific adviser (GCSA), within the Department for Innovation, Universities and Skills (DIUS); he was appointed on 1 January 2008. This department reports to the prime minister and Cabinet and is responsible for the quality of scientific advice to them on scientific and science policy issues (Cabinet Office, 1993).

Foresight is well placed within the various government departments as program members have the opportunity to work across all relevant government funding bodies. In one of their recent reports, the authors analyze their thinking to include questions such as "What will be the return on investment for Foresight funding?"[3] When it comes to the country-wide research agenda, they pose questions that then become the subject of debate and comment. Our discussion here concerns novel ideas for the manufacturing industry which can then be generalized and utilized by other industries. With that in mind, the Foresight website includes questions such as the following:

- Should academic research turn its attention to materials that are difficult to handle, such as powders and gels, and to processes with complex phase relationships?
- Is there enough academic research interest in process industries outside of the conventional petrochemicals sector?
- Is there enough cross fertilization between research into product properties and processing?
- Where are the revolutionary ideas in processing? Who can turn carbon dioxide and water into useful chemicals?

These types of questions are designed to help to shape current and future government-funded technology initiatives that can be developed as part of the green agenda. Of course, the green agenda is long term: for example, some of its goals have to be met within the next few years, and some need to be realized by 2050. So among all the other issues is the need to foster a culture change in how technology can be developed or modified for the mid- to longer term, which does not necessarily answer some of the more immediate and pressing business concerns.

Indeed, they further comment,

- Green technology combines the concepts of green chemistry and clean technology, and its aim is to minimize waste and emissions during manufacture, rather than try to eliminate them afterward. The goal is maximum efficiency of output with minimum energy input. Ideally, it would use renewable or recycled feedstock.

- Green technology embraces industrial ecology where the aim is to link one process to another and so produce an integrated industrial estate that conserves energy and resources with virtually no environmental impact. It requires both novel chemistry and innovative chemical engineering design, and a rethink along the lines of leasing chemicals rather than selling them.

- The UK chemicals sector has made great strides over the last decade to improve its energy efficiency and reduce its impact on the environment. These improvements have been achieved mainly by the application of end-of-pipe solutions to existing processes. The next step is to embrace green technology.

- It is also starting to happen in Europe, where carmakers emit 160,000 tonnes of solvents per year from their paint-finishing operations. The firm Industrial Copolymers Ltd, for example, who were winners of the 2000 Green Chemistry Industrial Award for small and medium firms, has found a way of replacing the solvent in polyurethane paint coatings.

- The estimated global market for environmental technology is circa $335 billion, equivalent to that of pharmaceuticals. It involves higher risk but offers a massive opportunity for the sector in terms of both enhanced manufacturing economics and gains from technology licensing. These benefits have not been clearly articulated; for example, the toolkit Life Cycle Analysis is not widely understood or applied. It requires a close marriage between chemistry and chemical engineering to achieve success and applies from the research stage right through to implementation.

- In U.S. industry, there is already a clearer recognition that green technology equates to increased profit, and the Presidential Awards for clean technology applications reinforce this message. Cargill Dow is building the world's first large-scale facility for making plastics from corn. The plant will become operational in 2002 and convert 370,000 tonnes of corn per year into 150,000 tonnes of polyactide (PLA) to be known as NatureWorks PLA or Ingeo Fibres.

- Movement along the evolutionary path of green technology will depend upon incentives to bring about change. However, in the main, the drivers are weak and the new approach challenges existing cultures in government, industry, and academia.

The challenge here is that the chemical-manufacturing industry covers many diverse products and manufacturing conditions and therefore would need to be broken down into manufacturing tasks. In terms of energy usage, the iron and steel industry uses large quantities of energy, followed by the chemical, food and drink, and paper industries. Furthermore, it may not be just the energy usage that is of prime interest. For example, some industries use vast quantities of pure water which needs to be obtained or processed from less pure sources. Additionally, each manufacturing sector may use a different subprocess, so trying to pick out patterns and make general comments is a challenge.

Arguably, the techniques and implementation strategy known as *process intensification* (PI) represent a significant advance on the previous incremental improvements that were made within the chemical engineering community over many years.

There is a lot of chemical jargon and some technical aspects to the implementation of PI that are covered in the Foresight reports but have no relevance to us here. However, there is an organization called Energy Transition that defines PI as follows:[4]

How can efficiency in the production chain and in the consumption phase of products be improved? This is the central question addressed by the Platform for Chain Efficiency. The Platform does not focus on incremental improvements, but on technology breakthroughs and system innovations that save large amounts of energy and materials—innovations such as Process Intensification.

What is Process Intensification? Many installations in the process industry use technology that has been fully optimized, but is almost 50 years old. However, new technologies are becoming available which change processes so fundamentally that a much smaller installation can do the same job. This new technology is known as Process Intensification (PI). The massive skyline that now characterizes the process industry may look totally different in the future. As part of this intensification, far-reaching savings are often possible in energy, CO_2 emissions and costs.

The website also makes comments concerning the additional advantages of PI:

- Requires lower investment and operational costs.
- Uses less energy.

- Needs less space.
- Needs less material for building the installations.
- Yields higher product quality.
- Provides improved safety.
- Offers flexibility: small, modular plants are possible.
- Makes restructuring of the production chain possible.

So having worked through the currently favored British government–sponsored initiatives concerning energy issues, we have arrived at PI. Where did it start, what are some implementations that have delivered on their objectives, and what common methodology can be teased out of the various implementations that might be relevant to other industries?

7.1 PROCESS INTENSIFICATION

7.1.1 History and Development

Arguably, Colin Ramshaw is known as the founding father of PI having pioneered gas and liquid mass transfer during his time at ICI. The term *process intensification* was coined in the 1980s. Ramshaw ignored the current design of the existing plant and started with a blank sheet of paper, then concentrated on the key issues or steps in the process. This approach led to the prospect of much smaller chemical plants, hence the term *intensification*. Ramshaw was subsequently appointed professor at the University of Newcastle upon Tyne and became one of the founders of a company specializing in PI solutions (Ramshaw, 1995).

Let's just have a quick look at one of the early processes that was intensified. This particular example has been published on Ramshaw's company website.[5]

In general, gases mix well anyway as do low viscosity liquids in thin films. Simple geometry teaches us that smaller, finer packing gives us more and more surface area so that would be the obvious way to go—a column with very fine packing with counter-current gas flow. However a liquid film running through a bed of fine material is problematic when the liquid film thickness is around the same as the clearance between the bits of packing. Liquid flow essentially stops and the column floods.

The key therefore is the thickness of the liquid film and what controls that. Fairly straightforward equations describe that and although most of the factors relate to the physical properties of the fluid, one is independent namely gravity. The higher the applied gravity the thinner the film and the smaller the packing can be. That gives a lot of mass transfer surface area for volume i.e. an intensified plant. In order to increase gravity the centripetal effect of rotating the packing was realised in a "high-g" machine that resulted in an order of magnitude reduction in size.

What is important about high-g mass transfer is the thinking process that led to the development, rather than the value of the device itself. That approach involved understanding the key variables that underpin mass transfer and then building a machine that provides the right conditions to achieve enhanced mass transfer.

In more general terms it is about understanding a process (a reaction, a crystallisation etc.) with sufficient depth so that the key rate controlling steps are understood and to then match that process to the right processor. This is all very much about recognising fundamental characteristics and matching those to appropriate machines.

Most, if not all, of the original examples were driven as capital cost reduction activities, particularly in the late 1970s. Of course, more recently other benefits such as enhanced intrinsic safety, reduced environmental impact, reduced energy consumption, and rapid plant response have all been identified. Indeed, the original ICI concept was described as "the reduction of process plant volumes by 2 to 3 orders of magnitude."

The so-called Higee technique,[6] invented by Professor Ramshaw, represented a revolutionary change in process plant size reduction. Bart Drinkenburg of the major chemical company DSM[7] commented on the difference between conventional distillation columns and those that have been designed using PI techniques as the "size of Big Ben, to a few metres in height."

At approximately the same time as the Higee, Tony Johnstone developed the compact heat exchanger in the form of the printed circuit or diffusion bonded unit. Johnstone's design is now recognized as the design basis of microreactors, where the fine channels promote both rapid heat and mass transfer, and give the unit a powerful multifunctional capability.

More recently, Professor Ramshaw, Professor David Reay (of David Reay & Associates, United Kingdom), and Dr. Adam Harvey (Process Intensification Group, Newcastle University, United Kingdom) have organized the Process Intensification Network (PIN). The PIN website contains a section concerning PIN membership and scope and also the PIN objectives.[8]

Membership of the Process Intensification Network is open to all interested parties, companies, research laboratories, academic institutions and individuals, regardless of size, provided that they can see benefits from participation in the Network, in particular leading to further development, use and exploitation of process intensification. PIN is a bridge between academia and industry. Its principal roles include technology transfer, education and promotion of awareness.

Current membership is about 350, 50% being from industry and the balance divided between research institutes and Universities. A substantial proportion of members are from overseas, predominantly from Continental Europe.

The objectives are listed as follows:

To maintain PIN as a forum where all those interested in the science, technology and application of Process Intensification can communicate effectively with others in the PI and related fields.

To create an environment, based upon meetings and other means of communication, where members can gain:

- a knowledge and understanding of the science and technology of PI
- an insight into the activities of the research community in the area of PI, including identification of centres of excellence in the field
- a knowledge of the key needs of user industries as they adapt to the benefits offered by PI
- facilities for transfer of PI technologies between sectors
- a good knowledge of, and opportunities to participate in, research, development and technology demonstrator programs, funded by consortia of members and/or national/international funding authorities
- a route to further education in the science & technology of PI and its applications

The benefits of PI have extended far beyond the reductions envisaged in the late 1970s to early 1980s. Indeed, there are some common themes between PI and lean thinking. For example, a PI implementation may lead to the following:

- Smaller inventory levels
- Ultra-short chemical residence times which facilitate just-in-time manufacturing
- Better product quality
- Lower waste levels usually by increased product yields
- Energy reduction—one of the current business drivers

Of course, one of the natural benefits of a PI implementation, the reduction in plant footprint and the drive to incorporate upstream and downstream manufacturing processes on the same site, is also common to lean thinking concepts, the corresponding business driver in lean thinking terminology being reduced motion. In the early years, there was only a sporadic uptake in PI techniques; however, over time the chemical industry, at least in the United Kingdom, has started to embrace the technology so much so that the BHR Group Ltd. (fluid engineering specialists) has now amassed over sixty staff years of experience in PI implementations.[9]

The academic world—at least in Europe—now offers courses and qualifications in this field. For example, the University of Newcastle upon Tyne has a Department of Process Intensification and Miniaturisation,[10] which is often known as the PIM group. On the same website, the PIM group describes its activities as

> playing a major international role in shaping the future of the processing industries through the development of novel approaches to equipment design and process synthesis with the aim of intensifying and miniaturising process plants and making them environmentally friendly and flexible in terms of manufacturing capabilities and providing rapid response to market demand.

It also describes its main research interests as follows:

- Microcellular polymers
- Biomass waste gasification
- Nanostructured microporous materials
- Spinning disc reactors
- Oil–water and gas–liquid separations
- Carrier-mediated separations and remediation
- Membrane process (electrofiltration)
- In vitro organs and tissue engineering

- Biotechnology
- Enzyme encapsulation and bioremediation
- Agglomeration and microencapsulation
- Detergent processing

The PIM group comprises full-time academics, postgraduate students, and visiting experts, and as one might expect it has short-term and longer ranging goals.

Cranfield University also offers a short course in PI, the publicity for which states, "This course aims to introduce delegates to the principles and techniques of PI, to enable them to identify the opportunities and apply PI principles in their companies."[11]

The website also lists their current topics for this course as follows:

- Understanding chemistry for PI application
- The importance of mixing
- High-intensity inline reactors
- Combined chemical reactor heat exchangers
- Microreactors
- Rotating fields
- Enhanced fields
- Intensified separation
- Methodologies for PI

Over time, this short training course will change content as the organizers keep the material fresh and up to date. The current offering shows the depth and breadth of the material on offer and is intended by way of an exemplar and not a lasting definitive guide to PI.

So what do we know now that we didn't at the start of this chapter? Well, the bottom line is that over the last thirty years, there has been a growing body of knowledge and expertise relating to a proven technology that has a broad take-up in many sectors of the chemical process industries. Additionally, there have been many instances of savings both in terms of money and in line with the green agenda. Of course, a PI implementation is not a quick fix. It is intended to be a longer term activity that will address some of the same issues that are also covered in lean thinking. Indeed, none of the green agenda issues are covered in lean thinking, which is a drawback.

7.1.2 Current Implementations

Creating lists of successful PI implementations will not add value, partly because there will be a need to fine-tune the PI technique used in two different applications since the reaction conditions will vary as will the thermal profile of the desired reaction. Another obvious drawback to any list is that there is bound to be a favorite reaction or technique that gets overlooked.

These caveats aside, some generic methodologies are worth mentioning (Semel, 1997), such as the following:

- Fuel cells
- Membrane reactions
- Membrane separations
- Oscillating flows in reactors
- Reactive distillation
- Reactive extraction

Some continuous reactors, including simple techniques such as jet devices, plug-flow pipe units, and static or other mixing devices, have been employed in the efficient production of toxic compounds. However, developing a new chemical route, changing the nature or composition of a catalyst, and changing reaction conditions do not qualify as PI implementations.

There are a range of chemicals now prepared on an industrial scale using PI methods and equipment, including:

- *Caro's acid*, which is prepared by reacting concentrated sulfuric acid with hydrogen peroxide, is now made on demand, whereas previously it was manufactured and stored.
- *Methyl acetate* manufacture was modified by Eastman Chemicals. The engineers were able to replace six distillation columns by one multifunctional distillation column.
- *Monobromo-benzaldehyde* was conventionally produced in batch reactors with a typical productivity of about 15.5 kg/m^3/hr. The equivalent PI continuous process has a productivity of about 34.5 kg/m^3/hr.
- *Phosphoric acid.* A U.S.-based plant now manufactures 2,000 tonnes of P_2O_5 per day, reducing power consumption by a third and producing lower environmental emissions using less process equipment.

- *Phosphorus oxychloride* manufacturing 500 tonnes per month by conventional means would require three reactors, the combined volume for which is 34 m^3. In a continuous process PI installation, 700 tonnes can be manufactured per month using a reactor volume of only 0.5 m^3. An additional benefit to this process is the reduced water used as a coolant.
- *Styrene-butadiene rubber* can now be manufactured in PI equipment occupying 25 m^2, compared with a building of about 400 m^2 previously (while still producing 2,000 kg/hr).
- *Thionyl chloride* can be made continuously using a loop-type reactor with a heat exchanger in the loop. Conventional productivity was about 10 kg/hr/m^3 in a batch reactor, whereas the current PI process can produce 340 kg/hr/m^3. A conventional glass-lined reactor might typically have been 18 m^3, whereas a modern PI reactor might be of the order of 0.5 m^3.
- *Vermiculite* is used in the fire protection and industrial insulation industries. The conventional production method has rotary furnaces for heating and reaction. A UK-based company replaced a set of three rotary furnaces of 1.5 tonnes/hr capacity with a single fluidized bed furnace of 1 meter diameter with a capacity of 2 tonnes/hr. This increased capacity using smaller equipment afforded lower maintenance costs and overall energy consumption. Payback was achieved in sixteen months, and eleven plants are now operational in Europe.

Given the anticipated plant volume reductions, the toxic and flammable inventories of intensified plant are correspondingly reduced, thereby making a major contribution to intrinsic safety.

But what of the benefits of implementing PI? Of course, different implementations will realize different benefits. Additionally, there is often a business benefit in keeping this type of information confidential within a company. Nevertheless, these figures are representative of the data in the public domain:

- Capital cost reduced by up to 60 percent
- Chemical impurity levels reduced by up to 99 percent
- Energy reduction up to 70 percent
- First-time yield up to 93 percent of the theoretical maximum

- Reactor volume reduction up to 99.8 percent of previous reactor volumes
- Substantial reduction in operating cost

While safer operating conditions are implicit, it is very difficult to quantify potential benefits.

So the techniques are growing in number as more chemical companies implement PI technologies and the academic community develops its scientific methodology. But what is the methodology, and how can these PI techniques be generalized to other industries?

7.1.3 Methodology

Of course, any snapshot of PI methodologies will change with time as new techniques advance the number and complexity of the many PI implementations. In their paper presented at the Second International Conference on Process Intensification in Practice, Green, Hearn, and Wood (1997) of the BHR Group Ltd.[12] suggested that the methodology at that time could be summarized as shown in Figure 7.1.

Although there are many individual process elements, there are only three basic steps to this methodology. They are as follows:

- Understand the process and the business case, and identify any pitfalls.
  - This first series of steps should be holistic in that they should look at the overall goal and not get tied to conventional thinking or equipment. Those individuals who are creative "lateral thinkers" would be good to engage in this process as it may involve some "blue sky thinking," to use some business jargon.
- Identify and generate the relevant PI equipment and process.
  - There have been many PI implementations throughout the chemical industry. It may be relatively easy to piggyback on someone else's thinking and equipment. However, if this methodology were to be translated to other industries, there may be few, if any, precedents. Initially, PI implementations may require PI expertise from the chemical engineering community acting as consultants.

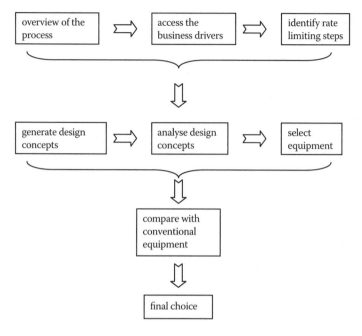

FIGURE 7.1
Process intensification (PI) implementation methodology.

- Undertake a reality check on the final choice.
- Part of the thinking in this phase of the methodology should be to determine that a PI implementation *will actually return cost benefits*. In other words, a high-risk PI implementation or one that will require long lead times or fabrication of novel process equipment may not be suitable for a product that returns modest margins but is manufactured in large quantities. In this case, a process replacement of conventional equipment may be the best alternative.

PI methodology will address many issues as described in this chapter. Equally, there are many business cases for not replacing a process but instead preventing the process from creating waste and/or defects. The waste could be in the form of process downtime rather than loss of materials, or bad product formation and rework which is also of importance. In this case, there is a technique that has been applied to many industries called *process verification*. This technique can be implemented in a very simple form or involve complex computer systems. Of course, the system that is implemented should be only as complex as needed to ensure that the process downtime is kept to a minimum *and* that the system is as easy to install and use as possible.

7.2 PROCESS VERIFICATION

The most simple process verification system is to hear, observe, or feel the process under consideration. In days gone by, steam railway engineers adopted the technique of wheel tapping, a process of hitting wheels one at a time with a hammer. The sound from the wheel enabled the attuned engineers to determine if there were problems with the wheel using the sound emitted from it after it had been struck. This simple technique required only one tool, some time, and a skilled engineer. I have had the privilege to know several engineers in the manufacturing process industry who were responsible for large complex plants. One used to go for a walk every few hours into the room containing a complex air-conditioning process. Tom could determine from the noise in the room if there was a problem with his part of the overall process. Engineers such as Tom are rare in that they have been allowed to develop their skills on one piece of plant and not moved from process to process. Over the years, Tom became attuned to his process and knew by the sound when things were awry. Similarly, sailing ship captains were often able to navigate ships by following a single star. There are many other examples of simple process verification systems.

At the opposite end of the data collection and analysis needs, there are industry sectors where simple systems might be impossible or hazardous to implement: for example, the production of explosives, radioactive materials, or some forms of energy for the power industry or for other uses such as the medical profession. These may require elements of process verification that cannot be accomplished by observation or some other simple technique. Another classic example is the manufacture of traditional silver halide–based photographic products. In this industry, turning on the lights to look at the process would in most cases ruin the photographic film or paper. In this case, a basic understanding of statistics and computing was found to be beneficial.

Figure 7.2 shows a process that is in control, at least for the parameter that is plotted. In this particular example, the data fall well within the

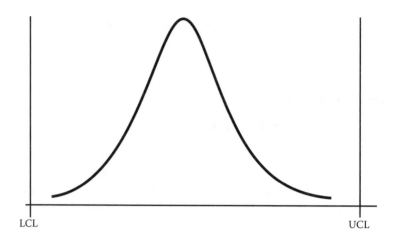

FIGURE 7.2
Variability consistent with a process that is in control.

upper control limit (UCL) and the *lower control limit* (LCL), so the plotted parameter is in control. One of the large photographic manufacturers produced its own process verification system that it was able to use to good effect in the production of photographic products.

Figure 7.3 shows a temperature profile of set point and measured value for a typical heating and cooling cycle. The two curves in Figure 7.3 are

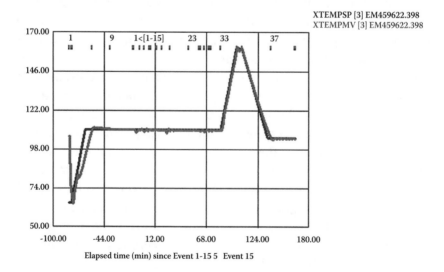

FIGURE 7.3
A measured value and set point temperature trace.

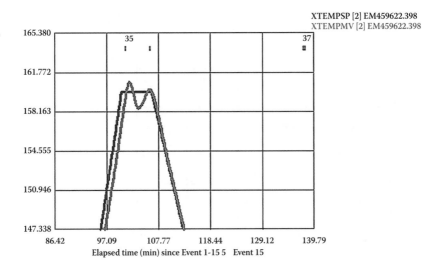

XTEMPSP [2] EM459622.398
XTEMPMV [2] EM459622.398

FIGURE 7.4
Temperature control trace at the elevated temperature.

the set point and measured value of a temperature ramp that is needed to melt out components and then heat treat this particular melt. The y-axis is the temperature of the melt, and the x-axis is the time from one of the process steps. This particular process verification system allows the user to horizontally align the curves based on particular events in the melting process. A more detailed view of the top of the heat ramp appears as Figure 7.4.

The left-hand curve of the pair is the set point, and the right-hand curve the measured value. It is clear from these traces that the temperature is not in control for the whole of the time when the melt should be at 160°F. For most of the melts made in this equipment, this temperature variance is not a problem. However, there is one product for which this level of variability causes product problems. This system highlighted the need to improve temperature control for this particular melting operation.

Figure 7.5 shows the difference between the set point and measured value for a period in the heat cycle of a melt which should be constant. In these circumstances, the difference between the two temperature profiles should be zero. A line corresponding to the desired aim of zero is plotted as the straight horizontal line.

The vertical axis represents the temperature of the difference between the curves and the horizontal axis the time at which the difference occurred

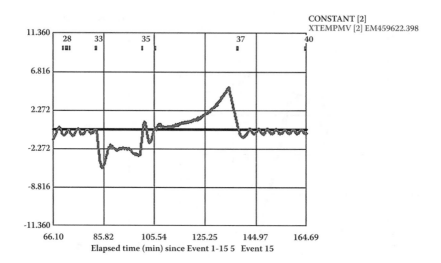

FIGURE 7.5
A difference plot between actual and expected temperatures.

from the start of melting. Both negative and positive excursions caused problems with this melt, the engineering solution for which was to change the temperature control algorithm. These melts are prepared in low lighting conditions, so this system of monitoring is essential to determine the performance of the process.

Figure 7.6 shows a Shewhart control chart of some of the temperature data for a part of the temperature cooling cycle. In the majority of cases,

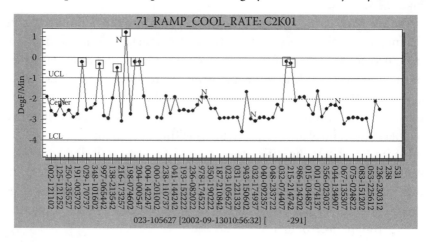

FIGURE 7.6
A Shewhart control chart of cooling rates.

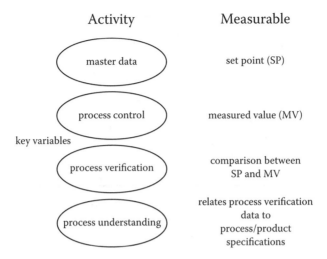

FIGURE 7.7
The various elements of a process verification strategy.

the data are within aim, but there are occasions when the temperature does not conform to calculated or historic limits. In these circumstances, there is a need to determine a means of ensuring that the melt is made in a consistent manner.

The above examples have been used to determine the health check of an actual plant in the chemical consumption industry. Of course, there will be variability in the heat treatment of melts that has no effect on product quality. Figure 7.7 shows how the process verification system became part of the data used in product understanding, and thus product quality.

The example shown above is part of a complex system that was developed over many years. In this case, there are known strategies for data capture, automated analysis, and alarming by exception those process data points that are collected. It must be stressed that this type of data would be of benefit for all sectors of the industry, but they are expensive to implement and maintain and may be a financially untenable option for some companies. Additionally, this process verification system required a high level of complexity because the number of products and melting cycles for intermediate stages was vast. At the opposite end of the product spectrum, the power stations which produce only one type of power will require a system of far less complexity. Indeed, systems for their use can be bought from specialist software houses.

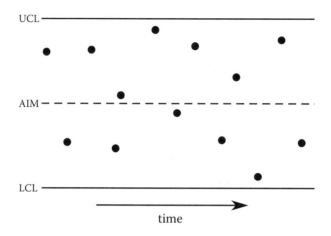

FIGURE 7.8
A typical control or Shewhart chart.

Figure 7.8 outlines a typical control chart that could relate to any process parameter; the circles are data points. In this case, regular readings of a process parameter with time produce a control chart where the upper and lower limits adequately describe the process data (see dots in Figure 7.8). This process is therefore deemed to be in control, at least for the parameter under discussion.

Figure 7.9 shows two control charts. One chart has been set at right angles to the other to better explain the effect of multivariant analysis. The horizontal data points suggest that this parameter is in control when taken in isolation of any other data. This is also the case for the control chart, which has been displayed vertically. When combined together, these data represent two process parameters where both variables have a relationship or dependency to each other and to the product. In this case, all of the data that are represented in the ellipsoid in the top left-hand corner of Figure 7.9 are in control. The data outside of the ellipsoid are out of control in multivariate space. Monitoring process data (the object of which is to ensure a stable process which will not affect product) therefore become an issue of multivariate proportions. It is nevertheless a vital part of the Six Sigma paradigm.

Some complex manufacturing processes may generate perhaps 3,000–4,000 analog signals and 1,000 digital signals on a second-by-second basis. Determining which parameters are product dependent and what relationship they have to each other is a nontrivial issue. Methods of automatic

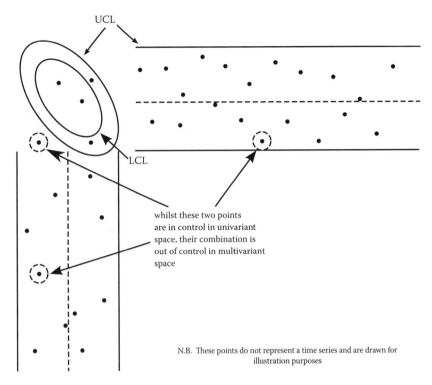

UCL

LCL

whilst these two points
are in control in univariant
space, their combination is
out of control in multivariant
space

N.B. These points do not represent a time series and are drawn for
illustration purposes

FIGURE 7.9
The relationship between two variables expressed as control charts.

data collection and analysis have been developed within some of the larger chemical companies. Some other database applications are commercially available. In any event, these data-logging and analysis techniques are vital to ensuring that there are no process-induced defects.

These data also serve other purposes. They can be used to determine a process failure, thus helping to reduce *mean time to recover* (MTTR). A regular inspection of the data can lead to better *mean time between failure* (MTBF). Reducing MTTR and increasing MTBF help a company to achieve better financial performance as does any other method of process understanding, even if it is used initially for fault detection.

Many U.S. and European companies have implemented some or all of these concepts to greater and lesser extents, with the degree to which these concepts are implemented varying within each organization. A company-wide Six Sigma implementation will require many thousands of hours from every company employee from the board members downward.

Of course, there are many matters to resolve with this approach. A fairly simple process involving few steps might be simple to determine. A more complex process may be difficult if not impossible. For example, a dye flow rate in a paint production process may directly relate to the eventual color of the paint—in which case this approach would work really well. Unfortunately, the effects of some process variables are additive, while in other cases the effects might be in opposition (as one variable goes up, another may come down). If you think that setting aims and limits was bad, try doing it in multivariate space!

So all we have to do is produce a product to aim specification, and we are home free! Well, actually no: there is much more to running a successful business than just paying attention to producing product to aim specification. This simple approach does not take into account inventory, cycle times, time to money, or indeed a host of other business metrics.

NOTES

1. http://www.foresight.gov.uk.
2. http://www.foresight.gov.uk/DTI/Pub 5203/2k/12/00/NP. URN 00/1276.
3. http://www.bis.gov.uk/foresight/about-us/criteria-for-selection.
4. http://www.creative-energy.org.
5. http://www.protensive.co.uk/pages/technologies/category/categoryid=overview.
6. http://chemelab.ucsd/higee/technology.html.
7. http://www.dsm.com/en_US/html/home/dsm_home.cgi.
8. http://www.pinetwork.org/about/pin.htm.
9. http://www.bhrgroup.co.uk/pi/index.htm.
10. http://www.ncl.ac.uk/pim/main.htm.
11. http://www.cranfield.ac.uk/soe/shortcourses/pse/page4789.jsp.
12. http://www.bhrgroup.co.uk/pi/index.htm.

REFERENCES

Cabinet Office. 1993. *Realising Our Potential: A Strategy for Science, Engineering and Technology* (CM 2250). London: HMSO.

Dyer, C. H., G. P. Hammond, C. I. Jones, and R. C. McKenna. 2008. Enabling Technologies for Industrial Energy Demand Management. *Energy Policy* 36, 4434–4443.

Green, A. J., S. Hearn, and M. Wood. 1997. Methodologies for process intensification. Paper presented at the Second International Conference on Process Intensification in Practice: Applications and Opportunities. Antwerp, Belgium: Author. ISBN 1-860-58093-9.

Ramshaw, C. 1995. The Incentive for Process Intensification. In C. Ramshaw, ed., *Process Intensification for the Chemical Industry*. London: Mechanical Engineering Publications. ISBN-10 0-852 98978-4.

Semel, J., ed. 1997. *Process Intensification in Practice: Applications and Opportunities*. London: Mechanical Engineering Publications. ISBN-10 1-860-58093-9.

8

Business Psychology Concepts

While the term *business psychology* is relatively recent, there have been many instances of psychologists working in various industries around the world. In addition, there has been an army of highly trained individuals at universities and institutes who have worked tirelessly to understand various aspects of team behavior. It would serve no purpose to mention all of these people by name; however, there are some notable individuals who will be referred to in this chapter.

The academic marriage of psychology, business practices, and indeed sociology can be traced back many years including the degree which started at Salford University in 1976. This particular degree, known as a bachelor of arts (with honors) in organizational analysis and industrial relations, was geared more to personnel management but covered some aspects now known as *business psychology*. Indeed, this degree was the first in personnel management.

The first European conference concerning business psychology took place in 1998 at the University of Westminster and was organized by Dr. (and now also Professor) Stephen Benton. He later went on to form the Business Psychology Centre[1] and then instigated the first European master of science degree in business psychology. Benton established a way of identifying and integrating individual differences in core competencies as defined in his business psychology model (see Figure 8.13 later), more of which in this chapter.

While I would hate to distill all of the accumulated knowledge from Professor Benton and other individuals into a single phrase, most of the issues and problems either studied or researched involve individuals, teams, or some form of change. Arguably, the model proposed by Gary Austin and inspired by the work of Dr. Elizabeth Kübler-Ross is useful in

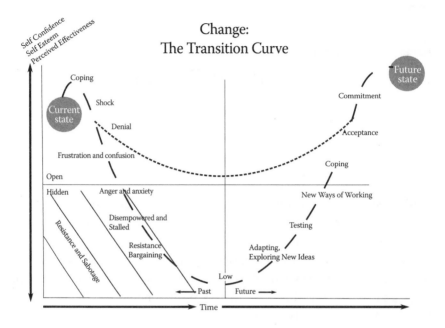

FIGURE 8.1

The culture change curve.

a variety of situations. Benton later modified Austin's model, a version of which appears in Figure 8.1.[2]

The curve in Figure 8.1 is interesting from all sorts of perspectives. Austin freely admits that he came across this curve when looking at some aspects of bereavement. As one might imagine, bereavement is perhaps the most extreme form of culture change. In this instance, there is a catastrophic change of events which results in a change of culture which one simply has no choice but to accept. Once Austin had made the relevant connection and proposed his initial annotated curve for use in the business context, Benton modified it as described here.

It is an interesting curve when applied to business for it can be used to rationalize some of the events in history. For example, the introduction of mechanized machinery during the English Industrial Revolution led to some workers damaging their machines with wooden shoes. These clogs were known as *sabots*, and the term for damaging a machine was *sabotage*, which is exactly the process that is described in the bottom left-hand quadrant.

Whilst the solid, annotated curve in Figure 8.1 describes the path followed by most of the workers, those initiating a company culture change follow the dashed curve as they have advanced warning and a longer preparation time to transition from the current to future states. Furthermore, some of them may even have been involved in the culture change process, for example blackbelts introducing lean thinking!

In this representation, the premise for a successful culture change is that there is a vision of the future state which is an improvement on the current state. This will be eventually measured by an increase in self-confidence, self-esteem, and/or perceived effectiveness.

The concept of a future-state vision can therefore not be overestimated. The new vision must be clear, concise, and unambiguous. It needs to be made available to all of the workers at *all* levels and should be the thrust of all communications between senior managers and workers during the course of the transition. It should be the common theme during briefings from the senior managers so that the rest of the workers have a consistent message. There is nothing worse than a future-state vision which appears to be developed during the transition stage.

Not all of the workers will necessarily experience all of the symptoms outlined on the solid curve in Figure 8.1. It will depend on each worker and his or her ability to understand the future-state vision and to commit to the new concepts. In general, however, there will be a degree of confusion, resistance bargaining, adapting, testing, and so on prior to full commitment to the new vision. Once committed, a worker should act as an advocate for the new state and help to "sell" the benefits.

Eventually, if properly implemented there will be a balance between the needs of the company, in this case the implementation of lean manufacturing techniques, and the needs of the individual (see Figure 8.2). The following few diagrams were developed jointly by Professor Benton and myself while we were writing a Six Sigma workshop a few years ago

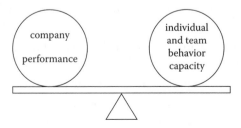

FIGURE 8.2
Future state.

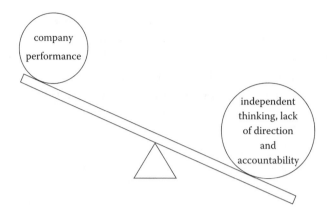

FIGURE 8.3
People issues dominate the culture.

and shows balance within a company which is actually the ideal platform for change.

If the needs of the people and the needs of the business are out of balance, there will be a tendency for the culture to drift in one of two ways. Figure 8.3 shows the balance being "upset" in favor of the people issues, which may manifest themselves as independent thinking, lack of strategic direction, and the like. In this case, company units become divorced from strategic aims as local norms and practices fuel "personal agendas" (Figure 8.3).

If the needs of the company are the dominant feature of the new culture, an imbalance occurs, perspectives become misaligned, and behaviors focus on inappropriate aims (Figure 8.4). The issue of concern is that a balance be struck between the needs of the company and those of employees. The company goals are often related to productivity and waste reduction, which are easier to achieve if the workforce is committed to the future of the company. Very rarely, however, can a worker on his or her own change a company for the better. It is more likely that a team working together within a project or functional group *will* make a difference. This has been described by Benton in terms of coherence.

Of course, there are experts who have studied culture change over many years. Some of these individuals have also studied the interaction of one culture with another. This aspect of cultural differences has become more important with the increasing globalization of large companies, and is relevant to the issues that have come to the fore as computer technology

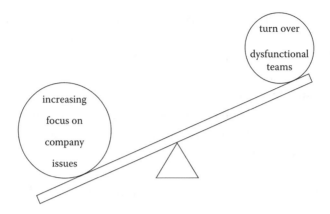

FIGURE 8.4
A culture in favor of the company goals.

allows workers to join and function in virtual teams. Virtual teams are interesting to work in as participants work toward common team goals while being resident in different locations or countries. I worked in a virtual team a few years ago and never saw two of the people in the team, even though we had worldwide conference calls every month for well over a year and produced a technical report documenting our conclusions.

8.1 CULTURE

8.1.1 National Cultures

So what's the big deal with culture, and why spend time discussing it and the impact of culture in business? In its broadest definition, *culture* can be expressed as mental programming; indeed, it has been defined as such by Geert Hofstede and Gert Jan Hofstede in their book *Cultures and Organisations: Software of the Mind* (Hofstede and Hofstede, 2005). Fairly early on in their book, the authors discuss the concept of patterns of feeling, thinking, and, perhaps more importantly, potential acting that have been learned throughout a person's lifetime.

Our mental programming is the total sum of our experiences garnered from our birth, family, school, neighborhood, community, and workplace environment. Of course, there are some individuals who react against their upbringing and leave their environments, perhaps to travel, set up a company, or make drastic changes to their lives in order to reject this programming in a conscious attempt to change themselves. There are others who embrace their surroundings, not always to the positive benefit of society. All too often, we read in newspapers or watch or listen to news broadcasts of issues relating to gangs and their exploits hitting the national consciousness. But in the main, we tend to live peaceably in our local communities and work for a local, or at least commutable, company.

The issues relating to culture are more complex than simply observing individuals and how they react to their own environment, since there are many layers to culture which help to make up the national identity. For example, there are layers relating to the following:

- The generational level
- The gender level
- The ethnic level
- The social class level
- The company level
- The national level
- The regional level

Sorting out all of the ramifications of all of these various cultural levels will add no value here. We can, however, gain an insight that is relevant to our discussion, from a study undertaken by the aforementioned authors involving workers who were employed in the same level of a company, but at offices around the world. International Business Machines (IBM) was the company that was chosen to help in the study as they had sufficiently large offices in many countries of the world. As one might imagine, there is little opportunity to validate these data using other companies as one would need to conduct the comparative research on similarly sized companies in the same countries—a difficult undertaking.

As one might imagine in a survey of this nature, the researchers took many months if not years to patiently work through each country and compile the results. For that reason, many questions were asked during the survey, the results of which are far too detailed to mention here. Let's just look at a facet of the end results that deals with a metric called the Power

Distance Index (PDI). In this part of the survey, covering fifty countries, workers were asked three questions covering the following three items:

- Nonmanagerial employees' perception that employees are afraid to disagree with their managers
- Subordinates' perception that their boss tends to make decisions in an autocratic, persuasive, or paternalistic way
- Subordinates' preference for anything but a consultative style of decision making in their boss, that is, their preference for an autocratic, persuasive, paternalistic, or democratic style

While the scores by country were published in Hofstede and Hofstede (2005), Geert Hofstede makes further reference to them in his book titled *Culture's Consequences* (Hofstede, 2001). Figure 8.5 shows some of Hofstede's data. Hofstede goes on to document the PDI scores which vary further by occupation as well as by country, demonstrating the levels of complexity. So now that we have a table of PDI scores, what do the data actually show? Triandis showed that male students with different PDI scores exhibited different behaviors—at least between the low and high PDI numbers. For example, Figure 8.6 is an extract from the data produced by Triandis et al. (1972).

Hofstede makes a further comment concerning the key differences between low and high PDI scores for family, school, and organization. Let's have a look at some of his data for the work organization in Figure 8.7.

Rank	Country	Actual Power Distance Index (PDI) Score
1	Malaysia	104
7	Arab countries	80
10/11	India	77
13	Singapore	74
15/16	France	68
33	Japan	54
38	United States	40
42/43	Great Britain	35
53	Austria	11

FIGURE 8.5
PDI scores by country.

Low Power Distance Index (PDI) Scores		High Power Distance Index (PDI) Scores
	Antecedents of "freedom"	
Respect for the individual's equality		Tact
		Servitude
		Money
	Consequents of "freedom"	
None given		Industrial production
		Disorderly society
		Wealth
	Antecedents of "wealth"	
Happiness		Inheritance
Knowledge		Ancestral property
Love		High interest charges
		Stinginess
		Crime
		Deceit
		Theft
	Consequents of "wealth"	
Satisfaction		Fear of thieves
Happiness		Arrogance
		Unhappiness

FIGURE 8.6
Low and high PDI scores.

Low PDI	High PDI
Decentralized decision structures, less concentration of authority.	Centralized decision structures, more concentration of authority.
Flat organization pyramids.	Tall organization pyramids.
The ideal boss is a resourceful democrat.	The ideal boss is a well-meaning autocrat.
Managers rely on personal experience and on subordinates.	Managers rely on formal rules.
Subordinates expect to be consulted.	Subordinates expect to be told.
Consultative leadership leads to satisfaction, performance, and productivity.	Authoritative leadership and close supervision lead to satisfaction, performance, and productivity.

FIGURE 8.7
Organizational low and high PDI scores.

There is also an expectation of more Nobel prizes in science per capita in countries with low PDI scores! Clearly these data and the social science behind the numbers comprise a cogent body of knowledge that is of benefit to individuals and organizations in understanding their environments. These types of studies are of particular benefit to members of multinational corporations who interact with their counterparts or indeed other coworkers from their own organization but who are indigenous to another country. As we have seen, national and organizational cultures each play a part in how these individuals approach and solve problems. Furthermore, a working knowledge of cultural differences will help international travelers within organizations when they meet their counterparts in other countries.

8.1.2 Cultural Diversity in Business

There are many dilemmas in management, and one relates to the cultural aspects of multinational corporations. Fons Trompenaars and Charles Hampden-Turner made the following observation in their book *Riding the Waves of Culture* (Trompenaars and Hampden-Turner, 2002):

> [T]he main dilemma which those who manage across cultures confront is the extent to which they should **centralise**, thereby imposing on foreign cultures rules and procedures that might affront them, or **decentralise**, thereby letting each culture go its own way, without having any centrally viable ideas about improvement since the 'better way' is local, not a global pathway. If you radically decentralise you have to ask whether the HQ can add value at all, or whether companies acting in several nations are worthwhile. (p. 189)

They go on to observe,

> [I]t is not a matter of how much to decentralise, but what to decentralise and what to keep at corporate HQ. A company that does not centralise information cannot cohere, but this does not mean that decisions cannot be made locally. (p. 190)

There is a need to respect cultural diversity and work toward a common understanding. Above all, we must keep to our own cultural values while working toward an awareness and respect for the cultures of others as we work toward transcultural competence.

There are some interesting problems, as one might expect, one of which relates to perspective. As we move toward another's perspective, that person may be moving toward our own. There is a danger that the two might pass like ships in the night! The dynamic is stronger if someone from a wealthy country is meeting someone from a country whose wealth may be less material. The takeaway message is the need to be aware of, and celebrate, cultural differences while being aware of the need to reconcile differences.

Trompenaars and Hampden-Turner (2002) suggest ten steps that are useful in achieving reconciliation (pp. 205–217):

1. The theory of complementarity
   - Most situations can be described as being on a continuum between rules and exceptions; the terms are therefore complementary.
2. Using humor
   - We become aware of dilemmas through humor.
3. Mapping out cultural space
   - Should we encounter a situation where cultural differences are less than easy to determine, one can map this cultural space with either interviews or questionnaires.
4. From nouns to present participles and processes
   - Not all nouns can be made into present participles, but if we know what we want, suitable words can be found.
5. Language and metalanguage
   - You must, for example, be able to see that a particular customer's request is outside of the universal rules set up by your company, yet be determined to qualify the existing rule or create a new rule based on this case.
6. Frames and contexts
   - The usefulness of thinking in frame and contexts is that the latter contain and constrain the "picture" or the "text" within them. The important thing to grasp is that context and text are reversible.
7. Sequencing
   - Values appear to clash and conflict when we assume that both must be expressed simultaneously. It is not possible to be right and wrong at the same time.

8. Wave and cycling
   - Have you ever stopped to wonder what happens to our values if we assume that they are wave forms?
9. Synergizing and virtuous circling
   - When two values work with one another, they are mutually facilitating and enhancing.
10. The double helix
    - This mental picture is used by Trompenaars as a summary of all of the above nine processes.

So, in summary, there are many aspects to culture, which are present in many forms in all situations. Some of the blackbelts that I know personally travel throughout Europe, providing lean thinking training. One of these contacts springs to mind. In a recent conversation that we had in a supermarket car park, one of them told me about his trips to some of his company's manufacturing sites that are scattered from England to the former Russian Baltic states. He is undergoing a tour of most of them delivering some training. At each site, his delivery and the material are the same. Given the cultural issues mentioned above, more details for which are contained in the references of this chapter, would workers in each of these countries understand what is needed of themselves to the same extent, given that they approach the subject from their own perspectives? If an audit were to take place, how effective would the training be perceived by course participants?

8.2 INTERPERSONAL DIFFERENCES

It is an old adage that the best asset within a company is its workforce. While it might be a hackneyed phrase, it remains a valid argument. Of course, it presumes that the company in question manufactures products or undertakes a service or services that the market is willing to buy. In the case when a company finds that revenues are going down because there is no market for its products, the first desperate action from the management team is usually to affect the workforce in some way. Such actions are usually designed to reduce costs affecting the workforce by a salary reduction, ban on overtime, reduced company pension contributions, or the like. However this might be achieved, the general mood within the organization in question is adversely impacted morale and motivation.

Ever since Jung developed his thinking concerning the self and others, various psychologists have attempted to use these concepts (and indeed other ideas) in the public arena in some way or another. Jung offered many studies of self and the influence of self on others throughout the course of his teaching. An understanding of introversion and extraversion led to a more complex model that has since been developed into a practical and commercial computer program which is now used extensively. Jung talked about three pairs of preferences:

- Introversion and extraversion (the way we react to experiences)
- Thinking and feeling (how we make decisions)
- Sensation and intuition (how we process information)

Each preference is an open-ended continuum.

The four preferences of thinking, feeling, sensation, and intuition were plotted on a circle by Jacobi from 1942 (see Figure 8.8) (1973, pp. 7–25). Those people with a combination of

- thinking and sensation tend to be objective, task focused, and thoughtful.
- sensation and feeling tend to be reflective and work toward a consensus.
- feeling and intuition tend to be action oriented, entertaining, outgoing, and sociable.
- intuition and thinking like working with others, are logical, and focus on the facts.

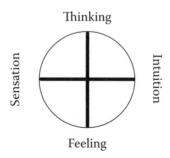

FIGURE 8.8
Jacobi's four preferences circle.

While there are many psychological tests using Jungian psychology, that developed by Insights Ltd.[3] has been proven to be accurate, and provides an in-depth analysis of an individual and that person's ability to work with others, or it did until quite recently! The developers have used tens of thousands of Insights Discovery Personal Reports to build up a highly reproducible model that has statistical validity in many countries, reflecting cultural diversity as discussed in this chapter.

This model uses a circle and colors to reflect the various combinations of introversion, extraversion, sensing, feeling, intuition, and so on. This not only provides a useful shorthand to the model but also allows the practitioners to create a common "language" which helps in many situations. As one might imagine, the basic circle is now much more complex than the early work of Jacobi (1942).

It is not the purpose or intent that this book should be a commercial for the Insights wheel or indeed the Insights profile or any of their other toolsets. However, this wheel has proved time and again to be of practical importance in helping with issues connected to interpersonnel relationships and team working. An individual is presented with twenty-five blocks of four questions per block, which he or she rates as *most like me*, *least like me*, and then a numerical score for the other two questions. The basic analysis using the Insights software produces a nineteen-page report covering such issues as the following:

- An overview which includes personal style, interacting with others, and decision making
- Key strengths and weaknesses
- Value to the team
- Communication
  - Effective communication
  - Barriers to effective communication
- Possible blind spots
- Difficult person
- Suggestions for development

This report is extremely useful for someone undertaking or undergoing counseling. It is of significant practical use when used in the team environment, and also provides a detailed profile of the individual. As with most activities, it has a project team composed of team members who have a range of attributes and use a diversity of thinking to solve problems. Having the opportunity to construct a project team from individuals who approach problems from different perspectives creates a project team that has a high degree of interaction and discussion, but is usually successful.

There is also further information in the report which examines how our personality changes under different circumstances. For example, a person at rest with no worries might be an extraverted thinker. Under stress, that person might exhibit more introverted thinking. Additionally, the report highlights the least likely of the four basic personality types that a person might exhibit, whatever the circumstances.

Furthermore, the model also recognizes that there are more than the four basic personality types of extraverted-thinking, extraverted-feeling, introverted-thinking, and introverted-feeling. For example, there are individuals who are neither introverted nor extraverted. Additionally, there are individuals who are neither thinkers nor feelers (in terms of the model). The model copes with this by listing eight basic types rather than the expected four, reflecting a wider variation of reality than other psychometric tests.

The first few pages of the feedback report cover the personality of the person, and additional pages concern interactions with others. It is an obvious thing to say, but we should always get on and be able to work with those individuals who think and act as we do. Placing team members on the same wheel will confirm which team members share the same quadrant and are likely to get along with each other. Similarly, such a plot will identify those team members who think in exact opposite ways. Now, on a good day we can all recognize strengths in each other, and while we are not always willing to admit it, there may be times when those individuals who think in the opposite way to us might be able to get a job done that we ourselves might struggle to complete. So we might have grudging respect. Some managers use this characteristic to their advantage and deliberately work with those individuals whose personality is opposite to their own as they can then receive a more rounded feedback on which to make decisions.

The rub comes when one works with or interacts with those people who are in adjacent quadrants to that of our own personality. Have you ever said of another individual that you "just don't know where they are coming from" and can't possibly imagine how they can think or feel as they do? Well, believe it or not, that other individual will be thinking the same of you! It's not right or wrong; it's just that we all have different personalities. Enlightened managers and leaders work with and celebrate different personalities so that they can create functioning teams with all of the various personality traits represented. Some managers have an instinct for this type of "people study," whereas others require help.

Even those managers with few inherent "people skills" will know not to create a team of people who only have the same personalities. This group might have a great time enjoying themselves; however, the project aims may not be achieved despite many hours of hard work on everyone's part. So the trick for the project manager is to draw from the collective strength of a diverse group while making sure that individual team members feel that their input is valuable to the team.

Without going into too much detail, this model helps solve practical problems of leadership. My first difficult situation as a team leader was to try to understand and deal with a problem within the team that I was asked to manage. Two of the team members had not spoken to each other for a few months, making life difficult for all of those people around them and indeed for the team that I managed.

I invited some business psychologists to write and deliver a one-day workshop which involved the Insights psychometric test. All of my team volunteered to complete the psychometric test and take part in discussions. They all followed through on their commitments and completed the questionnaire. It became obvious to all concerned that the two individuals in question were able to understand that they would never willingly wish to meet in a social setting because they approached life in very different ways. More significantly, they were able to appreciate that there are language and phraseology that helped them to communicate with each other. From the completion of the workshop, things became easier for both myself and the team. Additionally, I was able to look at all of the team members and pair up those team members with complementary personalities and, in one case, opposite strengths. It worked.

This psychometric test is not the only test that could be used. Indeed, many are available, each concentrating on a different aspect of the overall issue. One provides a descriptor that closest matches your personality such as *shaper, chair, company worker*, and so on (Belbin's team role analysis). While this test is extremely useful in ensuring that a team composition contains a variety of personality types, this particular test has not concentrated on interactions between the various personality categories. One of the other tests explores an individual's personality with a view to proposing possible careers. Others explore different issues most often used in interview situations. A blend of these other psychometric tests will provide some very useful data but may be less practical in the team setting. They, too, can have difficulties achieving acceptable and proven norm referencing across cultures.

Where two or more people are gathered together for a common purpose, an understanding of one's own personality and those of the other people in the group is a distinct advantage. On its own, however, there are still many challenges.

I used to think that the Insights tool was as good as it gets. I now know better, as there is a new kid on the block in the shape of a new model by a company called Luminalearning. In my opinion, the Lumina Spark Portrait much better reflects an individual's preferences than does any other model, Insights included. The training I undertook, along with the report, better reflects my personality traits compared with the last time I received an Insights Discovery Personal Report. Luminalearning is a new company, and I therefore do not want to go overboard and only recommend Lumina Spark Portraits. However, if their early promise pans out into concrete achievement over the next few years, I will definitely recommend Luminalearning products over Insights tools. Time will tell.

8.3 TEAMSHIP

Theories of teams and the process that a successful team should follow have been reported in the literature for many years. Arguably, the contribution from Tjosvold[4] in 1994, modified and improved by Benton in 2001, and incidentally now taught at an advanced-degree level in business psychology degree programs,[5] will prove instructive in determining what makes a team successful. Figure 8.9 outlines the Tjosvold model in its original form (Tjosvold and Tjosvold, 1992).

Benton modified the Tjosvold model in 2001 by changing the focus to Team Coherence and the Drivers for Alignment and Mobilising (Figure 8.10). Benton discusses his thinking concerning his modifications to Tjosvold's model in using an argument concerning coherence. His business psychology model provides a framework for identifying and utilizing differences, building coherence and clarity of information exchange (see Figure 8.11).

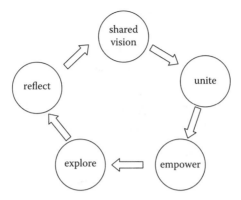

FIGURE 8.9
The Tjosvold model.

Benton describes the scenario of workers influencing a team and the team influencing the organization. This makes intuitive sense as there has always been the need to develop interpersonal skills which then go on to affect team behaviors and then the organization. He also proposed an alternative form of the Tjosvold model, where the emphasis concerns interpersonal coherence for the individual rather than the team. In this case, competencies are needed to provide the Tjosvold framework with an internal dynamic driving positive resolution (see Figure 8.12).

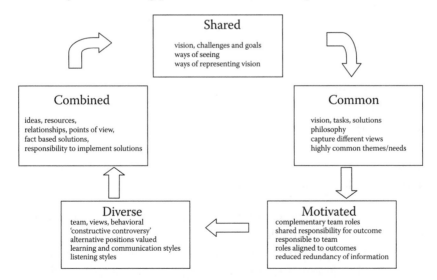

FIGURE 8.10
Benton's modified Tjosvold model.

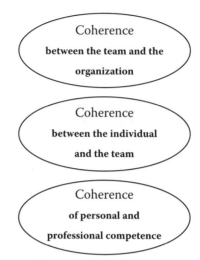

FIGURE 8.11
Project teams and coherence.

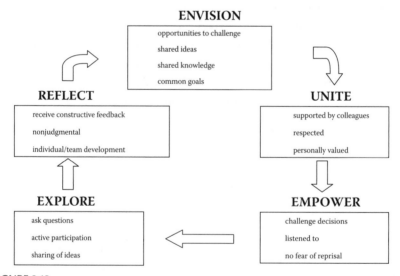

FIGURE 8.12
Interpersonal coherence for the individual.

This is an interesting development because it helps to tie together the thoughts from our earlier discussion concerning interpersonal skills and Jungian psychology with the Tjosvold concepts. Not only does it act as a means of bringing the relevant thinking together, but also it provides some guidance to the would-be team members as they strive toward a better understanding of teamship.

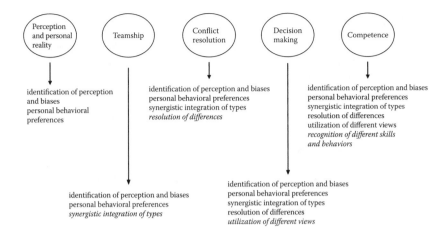

FIGURE 8.13
Benton's business psychology model.

Benton also proposed his own business psychology model in 2001 (Figure 8.13) further exploring the concept of teamship. This logical blend of skills and attributes combines to form a powerful model for team effectiveness. As we have seen in Chapter 6, there is some soft skills training in every blackbelt training course. However, the training concerning teams and team effectiveness is not as in depth as current thinking might require. Benton explained his thinking regarding this model as "the expansion of personal and interpersonal competencies as a basis for improving core organisational competence" (2010, personal communication). I would suggest that some or all of this knowledge could form the basis of additional modules within the lean thinking methodology and training, which will be discussed in Chapter 10.

NOTES

1. http://businesspsychologycentre.co.uk/buspsy.htm.
2. http://www.circleindigo.com/index.html.
3. http://www.insights.com/index.aspx.
4. http://www.ln.edu.hk/.
5. http:/courses.westminster.ac.uk/mSc-Business-Psychology-DO9FPBPY.aspx.

REFERENCES

Hofstede, G. 2001. *Culture's Consequences*. London: Sage Publications. ISBN-10 0-8039-7323-3.

Hofstede, G., and G. J. Hofstede. 2005. *Cultures and Organisations: Software of the Mind*. Maidenhead, UK: McGraw-Hill. ISBN-10 0-071-43959-5.

Jacobi, J. 1973. *The Psychology of C. G. Jung*. New Haven, CT: Yale University Press. ISBN-10: 0300016743.

Tjosvold, D. R., and M. M. Tjosvold. 1992. *Leading the Team Organisation: How to Create an Enduring Competitive Advantage*. Lanham, MD: Lexington Books. ISBN-13: 978-0669279726.

Triandis, H. C., V. Vassiliou, G. Vassiliou, Tanaka, and A. V. Shanmugam. 1972. *The Analysis of Subjective Culture*. New York: John Wiley. ISBN-13: 998-0471889052.

Trompenaars, F., and C. Hampden-Turner. 2002. *Riding the Waves of Culture*. London: Nicholas Brealey Publishing. ISBN-10 1-857-88176-1.

9

Changing Attitudes

There is a rock band called Queen[1] that may be known to people of a certain age. Although still active, the group came to prominence in the 1970s and 1980s, producing some outstanding music that filled stadiums around the world. Their lead singer at that time, Freddie Mercury, sadly died of AIDS on 24 November 1991. One of the songs written by Mercury, knowing that he was dying of AIDS, appeared on the *Innuendo* album and was called "These Are the Days of Our Lives"; it was released in 1991. The first few lines of the song are as follows:

> Sometimes I get to feelin'
> I was back in the old days—long ago
> When we were kids, when we were young
> Things seemed so perfect—you know?

While this particular song has been described as Freddie saying goodbye to his fans, this was also the time that corporate business rules were beginning to change. In many ways, the lyrics also reflect the passing of the "old" business models. The period from the late 1980s to now has seen many changes, not least of which have been the relentless advances in technology and business acumen of the Eastern companies and subsequent decline of Western corporations. The World Wide Web, in its infancy in those bygone days, is now a force to be reckoned with, particularly in the music industry. While CDs and vinyl records are still manufactured, modern music listeners are increasingly downloading favorite tracks. This trend is continuing in many business sectors. In addition, whole new businesses have been started which would not have been possible without the internet.

In his book *Japan's Global Reach* (Emmott, 1992), Emmott commented,

[B]etween January 1980 and January 1990 Japanese firms made overseas direct investments totalling $280 billion, equivalent to buying or building the entire economy of Australia, India or Brazil.... (p. 8)

[I]n late 1990, the Japan External Trade Organisation counted a total of 9,560 Japanese-affiliated companies with operations in the United States and Canada. Perhaps a third of those have factories or some sort of fully fledged branch office. (p. 5)

On the fortunes of Unipart, Emmott made the following observations:

Structures, techniques, systems; it all sounds rather mechanical. But that is misleading. In practice, the important task at Unipart was to change attitudes and to keep them changed. This did not only mean changing workers' attitude to their work; just as crucially, it also meant convincing workers that the company's attitude to them had changed too. Beginning in 1987, Unipart has been drumming this message home through four separate, but overlapping, 'people programs' all designed to show that people matter to the company and that the company's future depends entirely on its people. (1992, p. 98)

It is interesting to look back on this period since it was in 1990 that I undertook my first management course. I was sent on a business awareness program by my then manager. This program was effectively in three parts. These parts involved were

- a week at the Manchester Business School,
- a week of in-house talks from senior managers, and
- a group project.

The week at the Manchester Business School brought us into contact with some of the school's foremost thinkers of their day. They lectured and discussed the best practices of the day, including:

- Competitive marketing
- Financial budgeting and reporting
- Financial decision making
- Operations management
- Organizational development

A modern-day course will doubtless cover similar issues, although with modern insights and techniques. The week of in-house training was essentially for course participants to gain exposure to various parts of the company and how those parts functioned and fit into the overall business. The twist so far as we were concerned was that the speakers managed those operational units. The managing director opened the week's proceedings, passing on to a variety of board members and senior managers.

The project that I was involved in concerned communications within a part of the company where most of the staff were on the road in either sales or marketing. The brief was to assess the efficacy of communication within the unit. Now remember that this was before mobile phones and the internet. Many of the meetings between my fellow team members and the unit staff took place at prearranged locations in motorway service stations and the like. I remember driving from London to Manchester one day just to speak to a warehouse operator and ask him how much he knew of the unit's mission statement (among a whole raft of questions!). Things have definitely changed in a short space of time.

What strikes me now about this particular business awareness course was that it was mainly concerned with the European business and seemed to mention the United States only because our parent company was based there. I remember when I joined the company in the early 1980s, we were told there was no such a thing as a competitor. The phrase we were expressly told to use when referring to other companies who were producing similar product lines was "other manufacturers." How that changed as we worked through the 1990s with market share continually being eroded and price squeezes shrinking margins!

While I attended many training courses throughout the 1990s, the next significant course was a corporation-wide European Management Development (EMD) Forum. This course was actually open to any manager worldwide who was sponsored by a senior manager. There were forty-five or so participants, the course running once every year to eighteen months. This was a significant course to be asked to attend since it took place at the International Institute for Management Development, often known simply as the IMD.[2] This Swiss business school is regularly posted as number one in the league table of European business schools.

My course took place in 2001, and the course content reflected the new business challenges facing all large companies, including:

- Industry and competition analysis
- Competitive marketing strategy
- Creativity, innovation, and knowledge management
- Rethinking demand and supply chain management
- Financing growth in the new economy
- E-commerce: unraveling the web
- Leading change
- Getting high performance from cultural diversity

Even the titles suggest the urgency and need to consider competitors, global markets, and e-commerce! There are many benefits in attending a well-respected and dynamic business school, even if it is just for a week. Not only do you have the opportunity to mix with other like-minded course participants, but also you receive exposure to some of the world's leading academics who support the business world, and therefore to cutting-edge thinking. The IMD actually takes things further, in that you also become an alumnus or alumna. The alumni database receives regular mail shots and emails sometimes pointing to webcam discussions, or sometimes to a modern critique of a business process. However the information on offer is presented, it is insightful and cutting edge.

There are other means of acquiring modern thinking than attending respected business schools or reading their mail shots. One alternative is to read the books or attend lectures or seminars presented by modern-day business gurus. Just as Deming and Juran, mentioned in Chapter 1, helped business leaders of their generation, so we have the likes of Tom Peters and others to help us understand the churn of modern-day business. I find some of the books by Peters to be not only well written but also well matched to my sense of humor. For example, in his book *Re-Imagine!* (Peters, 2003), he offers twenty ways to self-destruct; the top five are as follows:

1. Establish a "sell-by" date for every business unit.
2. "Buy" research and development. Pay so much money that you're forced to make the most of the acquisition!
3. Recruit the world's best … and pay the world's best compensation.
4. Change the top executive's assignments every thirty-six months.
5. Start a huge venture capital fund.

To counteract the tongue-in-cheek cynicism of this list, he also provides a list of his top fifty ideas, some of which are as follows:

- Leaders create opportunities.
- Leaders say, "I don't know."
- Leaders are rarely the best performers.
- Leaders are talent developers (type I leadership).
- Leaders are visionaries (type II leadership).
- Leaders are "profit mechanics" (type III leadership).
- Leaders understand that … it all depends!
- Leaders thrive on paradox.
- Leaders love the mess.
- Leaders do!

There are many more.

During the course of his book, Peters also makes a comment about the Forbes 100 list that started in 1917. Let's have a look at the data and then the comment from Peters.

- Of the one hundred companies listed in 1917, sixty-one had failed by 1990.
- Of the thirty-nine survivors, only eighteen still numbered among the Forbes top 100 of 1987.
- These eighteen survivors had underperformed the stock market by 20 percent between 1917 and 1987.
- Just two companies had outperformed the market over the same time period (GE and Eastman Kodak).
- Of the two who outperformed the market up until 1990, Eastman Kodak has now collapsed, leaving just GE from the 1917 list.

Here is Peters' commentary, which is a version based on that of Professor Clayton Christensen of the Harvard Business School:

Good management was the most powerful reason leading firms failed to stay atop their industries. Precisely because these giant, bureaucratic firms listened to their giant, bureaucratic, largest customers, invested aggressively in marginally innovative technologies that would provide their giant bureaucratic customers more and better products of the sort that they already had and therefore wanted more of, and because they carefully studied market trends, which always say 'Do more of what you're already doing with a micro-twist or two', and systematically allocated investment capital to innovations that promised the best returns, which are always the most conservative innovations, they lost the positions of leadership. (2003, p. 35)

It's not all doom and gloom, for there are many companies from the early 1900s who have gone from strength to strength. Let's have a quick look at the early history of Harley Davidson,[3] reproduced from their website.

1901   William S. Harley, age 21, completes a blueprint drawing of an engine designed to fit into a bicycle

1903   William S. Harley and Arthur Davidson make available to the public the first production Harley-Davidson® motorcycle. The bike was built to be a racer, with a 3-1/8 inch bore and 3-1/2 inch stroke. The factory in which they worked was a 10 × 15 foot wooden shed with the words 'Harley-Davidson Motor Company' crudely scrawled on the door. Arthur's brother Walter later joins their efforts.

Henry Meyer of Milwaukee, a schoolyard pal of William S. Harley and Arthur Davidson, buys one of the 1903 models directly from the founders.

1904   The first Harley-Davidson Dealer, C.H. Lang of Chicago, IL, opens for business and sells one of the first three production Harley-Davidson motorcycles ever made

So what makes Harley Davidson tick? One of their senior managers once said, "What we sell is the ability for a 43 year old accountant to dress in black leather, ride though small towns and have people be afraid of him."

In other words, it's the *experience* of riding a Harley Davidson that appeals to many men of a certain age. All that managers at Harley Davidson are doing is to cater to the baby boomer generation while also attracting younger clients. To that end, senior members of the Harley Davidson management team have attended training activities at Disney University.[4] Disney University was set up by the Disney Corporation to train Disney employees. They posted the following on their website:

At the Walt Disney World Resort, we have three distinct programs for which international candidates may be eligible: the Cultural Representative Program, the International College Program and the Australia/New Zealand Work Experience.

Cultural Representatives share their culture and customs with our Guests while representing one of the countries or regions we have recreated in look and feel at the Walt Disney World Resort.

Participants in our International College Programs are immersed into a work-integrated learning program at the Walt Disney World Resort, and must be enrolled in an accredited university or a recent university graduate.

The Disney Corporation, better than most organizations worldwide, understands the concept of an "experience." For many years now, they have honed their skills and trained their staff to offer some unique facilities in which the young at heart can enjoy themselves. Creating products in other business sectors that also provide an *experience* might seem in some ways par for the course, in other words the rationale for that company's existence. In practice, far too many companies lose their way in delivering what the customers want in a way that is profitable to the company. The management team at the Harley Davidson Corporation has learned how to achieve total customer satisfaction. Good for them!

9.1 NEW PRODUCT DEVELOPMENT

In the mid-1990s, the magazine *BusinessWeek*[5] observed that "the new product battleground is a scene of a[n] awful carnage.... [O]f the 11,000 new products launched by 77 manufacturing, service and consumer-product companies, only a little more than half were still on the market five years later."

One might think that the challenge is therefore to bring products to market faster than the rate of old product withdrawal. Actually, anyone can do that and they probably are! What really needs to happen is that your organization releases products within the research, development, and product launch cycle of your competitors (i.e., quicker)! Keeping up with the competition was not and is not the answer to growth; beating the competition to market became the name of the goal. Of course, not only is there a need to deliver new products within the cycle time of your competition, but also there is a need to make products that are error free. In their book *World Class New Product Development* (Dimancescu and Dwenger, 1996), Dimancescu and Dwenger grouped problems of new product research and development (R&D) cycles and effectiveness (still relevant today) thus:

Common Problems across Companies

1. customer needs not well defined or understood
   - in seven out of ten failure cases, user's needs were not carefully gathered by the development team. This kind of lapse leads inexorably to the wrong designs and lacklustre market acceptance
2. errors found too late
   - the best companies[6] have a design error curve that peaks early with few, if any, errors occurring at the time of production
3. management by interference
   - this may have worked in rigid, heavily bureaucratic organisations, but it does not work where there are far more outsourcing partners and a demanding customer environment. If, however, teams are sure that their decisions will be honoured and supported, projects can be managed with greater flexibility
4. too many projects
   - project overload is one of the first symptoms of senior management abdication of its responsibility
5. burnout
   - the learning this team experienced was neither recorded nor passed on; knowledge accumulated at great personal and company cost simply dissipated
6. poor communication
   - it is the system of management that breeds what goes wrong much more than the inherent complexity of the product systems (pp. 8–17)

I have tried to use the motor industry for most of my examples, so let's have a look at a product design "failure" from GM that was actually reported in the *New York Times*[7] on 29 December 1992:

A survey by the New York Times of new car models in 1992 revealed that GM had 61 models with two keys, one for the door and another for the ignition. All Japanese models and most Ford and Chrysler cars had one key for both functions. Why the difference? GM's car doors were produced at one plant. The steering wheels with ignitions at another. A company engineer acknowledged on the condition of protecting his anonymity that it was an organisational nightmare to think of coordinating doors and steering wheels.

The *New York Times* made a further comment on 7 March 1993: "GM has stumbled because its system for developing new models is slow and cumbersome.... They have not been turning the millions of bits of information about what the customers prefer into the features and performance characteristics that attract buyers." As we have seen recently, the situation at GM may not have improved since 1993 as they recently filed for bankruptcy protection.

Further in their book, Dimancescu and Dwenger (1996, pp. 67–79) redressed the balance by also documenting positive attributes about teams.

Important Teaming Characteristics

1. a senior team of line executives is accountable to the CEO or executive committee for design, objectives and monitoring of the process
   - the strategic process and its toolkit of preferred methods and techniques constitute a corporate asset from which a company can exact substantial advantage over less-well-managed competitors
2. project or program core and extended delivery teams implement the process and are accountable to the senior team for execution and results
   - an emphasis on competencies is one of the most important breakthroughs in moving away from functional chimneys
3. the charter spells out the project or program constraints and empowers a team to act around explicit expectations
   - the creation of a charter requires both clarity of expectations and precision of objectives from management
4. concurrency redefines the timing and the quality of teaming relationships
   - improved communication is a vital adjunct of concurrently managed activity
5. collocation (real and virtual) maximizes the opportunity for face to face contact to take place
   - Chrysler's move away from housing specialised skills in dispersed buildings may be the most dramatic example of large scale investment in collocation at a single site; more than 700 collocated individuals work on a single platform

6. training in group skills and information sharing is essential for individuals to interact successfully in teams

- Ford applied a similar philosophy when it embraced the quality movement early in the 1980s. CEO Donald Petersen's attendance in a basic statistical control class signalled to the whole company that this was a training process of strategic importance to the firm

Let's just take a minute or two to discuss *silos*, which in this context refer to compartmentalized thinking to protect turf, control budgets, command and control, and so on. There's a lot of it about. Unfortunately, when one hand doesn't know what the other is doing, there can be all manner of activities that produce very little. Well, actually that's not true: the research folks produce many reports and ideas, but it's just that they are not turned into products that appear in the markets and sell in large numbers. All too often, you see that business and R&D management have few points of common reference, often talking totally different languages.

That's not too hard to understand. The R&D culture considers long-term projects that may run into years. The personnel often have educational and cultural norms involving totally different experiences compared with, for example, technologists in a manufacturing department. I spent three assignments, each longer than a year, in various manufacturing departments and three in R&D posts over a twenty-three-year period. There is a remarkable difference in attitudes, experiences, and workplace timescales. For example, the product technologist is concerned about the last batch of production and the current state of the manufacturing process. Indeed, the technologist's office may have a light that signals the process is manufacturing product (as opposed to a scheduled downtime for product changes, maintenance, etc.). Product technologists live on adrenalin, or at least they can when things don't go quite to plan. Additionally, many of the manufacturing technologists that I came across left the R&D organization simply because they preferred the immediate feedback and excitement of real-time manufacturing.

Managing organizations to produce products to ever shorter deadlines and increased product quality is not easy. Rousell, Saad, and Erickson (1991) refer to this process as third-generation R&D. At the time of writing their book, these three individuals were senior executives of an international management and technology consulting firm. Between them, they had many years of experience across all of the continents.

In their book, they comment,

> More than ever, corporate executives and general business managers are troubled by the question: What are we getting for our investment in R&D? Underlying this question is business managers' pronounced feeling that they are spending much and that the return may not be commensurate with the spending. In the worst case, top management views its R&D function as a black box—the input is money, the activities within the box are not comprehensible, and the output is uncertain but never enough.

The basic principle is to engage with all of the other divisions within the organization so that the R&D strategic direction has input and buy-in from all of the other divisions. In this way, there are no surprises from the business units when a new product suddenly arrives from the manufacturing community or that the request for a sales promotion from the business units is planned.

In many organizations, this type of commitment from and input by all of the participating departments is a way of life. There are a few major companies, however, where there is still room for improvement.

Of course, there are companies where there is, or at least there was, cooperation between all of the participating departments of R&D, manufacturing, business unit, and marketing, and yet still the company failed to deliver on customer wants. The recent problems with the new products from Eastman Kodak might fall into this category. For many years, the senior managers knew how to create digital products. Yet the margins from digital products are less favorable than those from traditional films and papers. So the digital product introduction and subsequent withdrawal of silver halide–based photographic products were potentially fraught with problems. History records that share price dropped from the $23–25 per share in 2004 to $2–3 in 2009. Considering that their market share in the 1980s was around 80 percent and the potential photospace market worth billions, it is a remarkable drop in fortunes for a once great company.

Perhaps as with other companies before them, the senior managers at Eastman Kodak would have benefited from a crystal ball able to show future markets and technology trends. Arguably, we might look to the Nobel laureates for inspiration since those individuals have proven track records in thinking strategically as well as out of the box. Arno Penzias, who was awarded the Nobel Prize for Physics in 1978 while working for AT&T Bell Laboratories,[8] made the following comments:

> Beset by competitors who didn't have research labs of their own to pay for, AT&T's leaders nonetheless did their best to provide for its 'crown jewel'. As one year followed another, I did my best to repay that trust by helping to turn some of our scientific 'gems' into profitable jewellery.
>
> And then, I did something that surprised everyone—myself included. I decided to swap my job for something entirely new, moving from the world's largest corporate R&D organisation to California's Silicon Valley, premier incubator of tiny start-up enterprises. In retrospect, I can point to a number of contributing factors—most notably obligatory retirement age, then only a few years away. While arbitrary, the notion behind an age cut-off for senior managers had much to recommend it. I couldn't (and still can't see) myself ever being happy without something challenging to work on. Since getting another management-related job seemed too much of the same thing, I hit upon the idea of turning what I had been enjoying most into a full time job: helping to shape new ideas, and bring them to practical fruition. The more I thought about it, the more attractive this plan for my post-retirement life became. So attractive, in fact, that I soon decided not to wait much longer to put it into place.
>
> Once decided upon, my transition proved surprisingly easy. At the suggestion of the then Bell Labs President, I soon took on a new job—one in which I was to report what I learned about Silicon Valley and its workings, to my Bell Labs colleagues. Accordingly, I arranged to sit in on presentations made by nascent start-up enterprises to venture capitalists. I felt right at home in short order.

9.2 SMART ORGANIZATIONS

In this case, as with many others, more of the same is not the answer. In their book *The Smart Organisation* (Matheson and Matheson, 1998), Davis Matheson and Jim Matheson describe decision quality as comprising the following elements, each linked to the other as links are joined in a chain.

- Meaningful reliable information
- Clear values and trade-offs

- Logically correct reasoning
- Commitment to action
- Appropriate frame
- Creative, doable alternatives

Matheson and Matheson conclude with the following comment:

[O]ur experience with these and other firms confirms the ability of smart principles, processes and practices to improve decision quality across the entire organisation. We have observed applications resulting in strategic choices that create millions to billions of new value. These large figures are dwarfed, however, by the value that could be created if all strategic decisions met the standards of decision quality. The principles of the smart organisation point the way. By becoming a smart organisation, every enterprise—including yours—can create this greater value. (1998, p. 261)

Having made the decision to decrease the time between product launches, it's a great idea to let the manufacturing department know. There are lots of reasons, but one that I want to spend a few minutes discussing is the extent to which lean thinking principles, specifically product change time reduction in the manufacturing department, have been introduced. I have spent some time as a product technologist in a manufacturing division, and I know from personal experience that talking the talk is not enough. Product change times need to decrease in line with the increase in the number of products in the marketplace.

Specifically, we had about eighty products that were being manufactured in a single facility. Some of these products were high volume and made every week. Some products were made in small quantities once or twice per year. Driven to their extreme, just-in-time theory and a pull manufacturing environment would have one product of each type made to order. In practice there were minimum quantities since we coated melts onto moving webs and needed to prepare minimum quantities of material in the melting area. Furthermore, the delivery lines were flushed out with water between products so that there was no cross contamination between them.

Nevertheless, it was possible to decrease batch sizes to a minimum. This drove product change time reduction, which itself generated start-up waste. So the operators needed to change procedures to accommodate the new scenario. We were then balancing inventory issues against start-up costs. Smart accountants can work on the math so that there is a financial as well as practical end point. Although a struggle, all needs can be met unless someone in the head office thinks that diversification is the key to generating business!

In some industries, diversification is undoubtedly a good idea in that there will be market segmentation, which can be great for business. There are other industries where one product is the norm. Power generation springs to mind. On a countrywide basis, there is usually a single specification for power often expressed in hertz, volts, and amps. A diversification strategy to produce two or more types of power will not go down too well with anyone. It's an extreme example to demonstrate that there are other industries where product change simply does not happen. For the majority of industries, the reality is somewhere in between.

Releasing timely products into the marketplace with all of the infrastructure bases of sales, advertising, and marketing personnel on board does not, however, always ensure a positive result. Competitor products may come from "left field," in other words from a technology shift or a global competitor not usually associated with your product lines. In the opening paragraph to their book *Strategic Technology Management* (Dussauge, Hart, and Ramanantsoa, 1992), Dussauge, Hart, and Ramanantsoa make the following statement:

> [D]ramatic technology changes in the 1970s almost wiped out the Swiss watch making industry. Yet technology was the key factor which made it possible for the few Swiss watch brands which survived to meet the challenge posed to them by their Japanese, Taiwanese and Hong Kong competitors. The development of electronic watches upset the well-established Swiss firms, making the bases of their technical excellence obsolete virtually overnight. Indeed, the precision of quartz watches could not be matched even by that of the best mechanical watches; furthermore, quartz watches were cheaper to manufacture than mechanical watches. Because they were unable to identify the threat they were faced with, many Swiss watch making companies were forced out of the market. However, ETA, a firm resulting from the merger of several ailing companies of this industry, managed to turn the threat into an opportunity and has experienced considerable growth since the beginning of the eighties. (p. 1)

The Swatch was the result.

Dussauge, Hart, and Ramanantsoa (1992) make a further comment concerning technology and globalization:

> [H]owever, as the impact of technology increases, the significance of local and regional differences is diminished. Indeed, the cost of technology, which is often largely comprised of fixed costs, requires that the latter be compensated by a wider market scope. Thus, technological evolution and the growing importance of technology in many industries encourage globalization of markets. Computers, automobiles and semi conductors are examples of industries where advances in technology are so costly that global sales are necessary in order to remain competitive. This globalization of markets due to technology is one particular form of the blending of distinct strategic segments resulting from technological evolution.

I guess that it is what we always thought, were taught, or just knew. Of course, email and website advances have led more companies to adopt a business strategy that is global. Some of the impacts of e-commerce will be discussed in Section 9.3.

Of course, globalization is not a new concept, for there are many examples in the finance industry, not least of which is the Rothschild banking industry founded by Mayer Anselm Rothschild in Frankfurt. He and his five sons established branches in Naples, London, Paris, and Vienna, with more following later. The transfer of funds between the original and newer branches became the norm.

Arguably, the Japanese automotive and electronics industries were the first to develop international manufacturing facilities. In his book *International Management Cross-Cultural Dimensions* (Mead, 2005), Richard Mead commented,

> [T]his globalisation of production meant that Japanese companies could produce and sell almost anything anywhere. Their success in pioneering the techniques meant that by the 1980s Japan seemed to be the only economic giant set on a course of continuous expansion. The United States was facing a crisis of archaic manufacturing plant. Having rebuilt its industrial base since the Second World War, European countries apparently enjoyed an advantage, but were hobbled by fuel shortages.
>
> In the 1970s readjustments in the U.S. manufacturing sector caused a shift in economic power from the traditional heavy industries to the development of new industries associated with information technologies. In

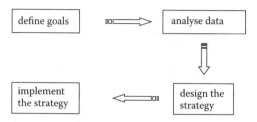

FIGURE 9.1
Global marketing strategy.

geographical terms, this meant a shift from towns like Detroit and Chicago to Silicon Valley. America's competitive edge in the development of the microchip technology helped to rebuild the U.S. economy. (p. 253)

Of course, what happened was the decline of one industry in one location and the growth of an unrelated industry in another location, the two industries having no relationship with each other. For a single business to grow global market share, there needs to be a global strategy just as there is for local growth. The necessary steps to creating this strategy are the old tried and trusted steps shown in Figure 9.1.

Operating in the global markets does not necessarily lead to an increase in the number of competitors or products. There are many instances where goods are either fabricated abroad or locally manufactured but from companies owned overseas.

In some cases, the goal may simply be to survive. Obviously, the longer term goal would be to grow the business to profitability. The financial goal of a strategy should be one of several parts to also include references to goals for the following:

- The market (maintain or gain on one's current position).
- Social considerations—these could be ethical or environmental.
- The organizational culture.
- The quality process.

There must be a review process that allows for changes in strategy as time and external influences become apparent. Long-term quality goals need to be a part of the overall process. Some major multinational corporations often create a unit just for personnel working on lean thinking, at least in the early days of lean thinking implementations. Often, the head of this unit is a proven blackbelt or a former employee of a Japanese corporation. In any event, implementing or improving the quality process ranks as one of the major strategic goals.

Multinational companies can have problems transplanting their strategic goals into the different cultures of their organization. There are cases where the culture of the foreigner cannot be readily adapted and is therefore not easily accepted. It is often beneficial in these circumstances to make incremental changes. Mead offers the following comments concerning the top positions in subsidiary companies:

> [W]hen should the company post a headquarters manager to run its foreign subsidiary, and when should it invest in a local manager?... If the headquarters needs control, it must be prepared to invest more. On the other hand, if headquarters wishes to decentralize control and focus on meeting local demand, it may decide not to incur the costs of an expatriate appointment and instead appoint a local manager ... when expatriates depend on locals, the latter transfer their skills to headquarters. To add insult to injury, expatriate managers from developed economies working in less-developed economies are usually paid far more than local managers....
>
> [T]he Japanese place great store on the advantages offered by an expatriate appointment. These include:
>
> - ease of control by headquarters
> - ease of spreading headquarters organisational culture in the subsidiary
> - greater control of local managerial and technical skill levels
> - greater protection of technology
> - common national culture and language (2005, pp. 358, 359)

The downside of this strategy is the opposite: for example, local managers have fewer development opportunities and the like.

9.3 E-COMMERCE

While there may have been a subtle difference between the terms *internet* and *intranet* in the past, most users use the two terms interchangeably. Wireless transmissions to mobile phones, personal digital assistants, and smart phones have helped to blur things even further. The internet can cause confusion, it can contain inaccuracies, and access can sometimes be slow, but it is without a doubt changing the way companies undertake business. The business world uses the internet in a variety of ways, including:

- Consumer shopping, sometimes referred to as *B2C*
- Business-to-business transactions, sometimes referred to as *B2B*
- Supporting sales and marketing through online catalogs or the like

Of course, there are also online libraries, academic institutions, and video or music libraries that may also be contacted by business users depending on the nature of their respective businesses. Sending emails is now considered routine, indeed essential, as is the ability to set up virtual team meetings.

In the early days of e-business, specifically between 1997 and 2000, more than 12,000 internet-related businesses were started with more than $100 billion. As we know from our history books, over 5,000 went out of business in the early 2000s. Since the *dotcom bust*, as it came to be called, many billions of dollars have been invested creating a stable e-business marketplace. Figures for 2007 of actual and estimated sales for B2B are $6,800 billion, and for B2C $240 billion. Just over ten years ago, the respective figures were $460 billion and less than $1 billion.

This rapidly changing marketplace led Schneider to comment,

[I]n the wake of the dot-com debacle that ended the first wave of electronic commerce, many business researchers analysed the efficacy of the business model approach and began to question the advisability of focussing great attention on a company's business model. One of the main critics, Harvard Business School Professor Michael Porter, argued that business models not only did not matter, they probably did not exist. It has become clear to many companies that copying or adapting someone else's business model is neither an easy nor a wise road map to success. Instead, companies should examine the elements of their business; that is they should identify business processes that they can streamline, enhance or replace with processes driven by internet technologies. (Schneider, 2007, p. 14)

There are well-documented advantages to both the seller and the buyer in e-commerce, not least of which is the ability to reduce sales and marketing costs and speed up cash flow. Indeed, there are some companies, such as Dell Computers, which obtain your money for a computer before they build it. This approach ensures that all of the computers sold by Dell contain state-of-the-art parts, and also they use the customer's money to fund their business. Efficiencies reduce overheads and in some cases drive down prices. Some products lend themselves well to internet sales. Downloading software, music and video files, and so on is much more efficient compared with ten years ago.

There are many web-based businesses which may not have "conventional" equivalents; additionally, there are businesses that have intentionally migrated to e-commerce. However, there are industries where technology is driving the business. We briefly discussed music downloads earlier. Book publishing is slowly migrating to the internet as books written for paper format also have electronic books created from the same typescripts.

The 2008–2009 subprime credit crunch has slowed down book printing for several reasons. As one might expect, there is now reluctance on the part of publishers to order large print runs and store the books in ware-houses in the hope of future sales. Also, all large-print runs are insured as they are costly. In these days of financial churn, some insurers will not deal with some businesses in the printing chain, thereby delaying, or in some cases preventing, printing.

Authors, myself included, now write books specifically for one market or the other. There is a different approach in writing an e-book, at least if the content is scientific (as mine tend to be). The reader is able to display only one or two pages at any one time, so my technique is to write information that can comfortably fit onto one or at most two pages. If it is spread out over two pages, it needs to be the two that are displayed at the same time. The reasons for this are twofold—ease of display and therefore ease of reading—and another reason is the ability for the reader to download just one page. I feel sure that books will follow the format of music in that the buyer will want to download just the information of interest. That may mean selling one page, one chapter, or indeed a whole book.

There is also the option of emailing out free chapters of novels and making the last chapter available to buy from a particular website. There are a variety of e-books formats, depending on the reader. The DNL Reader format is self-contained. A publisher can set up the book so that the potential buyer can read a few pages prior to registration. Once registered, the reader can then look at a predetermined number of pages before payment is required. All formats allow the book to be printed and can restrict the number of printed copies. There is even the option of sending e-books from a library. Prior to downloading from an e-library, the e-book file is programmed with the number of days that the file remains active. After this predetermined number of days, the file closes down and will not reopen.

Another advantage to the author from an e-book is the potential for sales from around the world. Publishers involved in the paper-based system may decide to release a particular book in one country only. I recently published a paperback book, and it appeared on a well-known, web-based bookseller within five days of publication. Furthermore, the French, German, and American sister sites also listed my book in the same time-scale. The nature of internet selling is such that as soon as an e-book is made available for download, all potential buyers from around the world are able to access and download the file. At the moment, e-book sales are restricted by the number of people who have access to e-book readers. Interestingly, the following comment was made recently by Dan Nystedt in his article concerning Computex Taipei 2009:[9]

> E-book shipments are taking off worldwide, says market researcher In-Stat, with worldwide shipments expected to grow from almost 1 million units in 2008 to close to 30 million units in 2013, due in part to the popularity sparked by Amazon's Kindle.

Only time will tell, however; the longer the world recession lasts, the greater will be the interest from authors in writing books for download for no other reason than the delays of having work published in paper format.

While most businesses use the web for selling, there are examples where the web has been used to create efficiencies during the manufacturing process. Schneider makes the following comment concerning an increase to supply chain efficiency at Boeing:

> [I]n 1997, production and scheduling errors required Boeing to shut down two entire assembly operations for several weeks, costing the company more than $1.5 billion. To prevent this from ever happening again, Boeing invested in a number of new information systems that increase production efficiency by providing planning and control over logistics in every element of its supply chain. Using electronic data interchange (EDI) and internet links, Boeing is working with suppliers so that they can provide exactly the right part or assembly at exactly the right time. Even before starting an airplane into production, Boeing makes the engineering specifications and

drawings available to its suppliers through secure internet connections. As work on the aircraft progresses, Boeing keeps every member of the supply chain continually informed of completion milestones achieved and necessary schedule changes.

By its second year of using the new system, Boeing had cut in half the time needed to complete individual assembly processes. It has realised similar reductions in part defect costs. The combined effects of these increased efficiencies are helping Boeing do a much better job of meeting its customers' needs. Instead of waiting 36 months for delivery, customers can now have their new planes in 10 to 12 months. (Schneider, 2007, p. 244)

This fascinating use of electronic data interchange (EDI) is very similar to the just-in-time and *kanban* elements of lean thinking. In this particular case, the supply chain process has been automated using computers and not lean principles. Dell Computer also uses EDI for supply chain management. Materials tracking is also much easier with EDI as all one needs are barcoded goods, or radio frequency tags, with appropriate readers. There are many EDI uses.

With all of these benefits, are there any negatives or at least downsides to e-commerce? There are a few, such as the following:

- Culture
- Nonverbal communication
- Trust

Culture is interesting in that it helps to determine laws and ethical standards. Of course, there are no geographical boundaries to the web. The legal community defines the relationship between geographic and legal boundaries in terms of the four elements of power, effects, legitimacy, and notice. Schneider comments,

[P]**ower** is a form of control over physical space and the people and objects that reside in that space, and is a defining characteristic of statehood. For laws to be effective, a government must be able to enforce them. Effective enforcement requires the power both to exercise physical control over residents and if necessary to impose sanctions....

Laws in the physical world are grounded in the relationship between physical proximity and the **effects**, or impact, of a person's behaviour.

Personal or corporate actions have stronger effects on people and things that are nearby than those that are far away....

[T]hus **legitimacy** is the idea that those subject to laws should have some role in formulating them....

[P]hysical boundaries are a convenient and effective way to announce the ending of one legal or cultural system and the beginning of another. The physical boundary, when crossed, provides **notice** that one set of rules has been replaced by a different set of rules. (Schneider, 2007, pp. 312–313)

A buyer in England unhappy with a web-based item shipped from the United States may find it difficult if not impossible to resort to the U.S. courts. There are some websites where it is almost impossible to determine where the website's owners are based. There are other issues to buying over the web, one of which is buying alcohol, which is regulated differently in different states in the United States. There is also the obvious problem of proof of age.

From a personal perspective, there are also the issues of intellectual property and copyright infringement. There are countless examples on a large number of websites where copyrighted pictures are openly displayed by unauthorized users. Additionally, it is very difficult to protect short articles that appear on websites and can be readily cut and pasted into other documents. The wider aspect of identity theft is a major concern for some people as is the large number of spam emails received from countless sources, suggesting that web crawlers routinely trawl for email addresses. It would be great if your email address could be unlisted and known only to those people you choose to inform!

While we are on the subject of communications, spoken and unspoken language is a feature of virtual team meetings that can be tricky. Over the years, I have participated in many voice-only conference calls with various sister plants around the world. Some involved a call in the early afternoon in London with participants from Melbourne, Australia, and Rochester, New York. The Americans arrived early and the Australians were sipping beer, it being often 10:00 or 11:00 p.m. their time. I have to say that there were a few classic calls when we were definitely divided by a common language. Subtle variations in words in one country were totally lost on another.

Voice-only calls with our colleagues in France and Mexico were sometimes challenging as we often found it difficult to understand each other. Humor doesn't always travel either!

It may sound as if I am against the opportunities that the web brings to new or even established businesses. Actually, the opposite is the case. I spend a large amount of my day on the web researching material for books. I have learned to be careful about websites that I trust, often using favorites such as the British Library, the Royal Society, the Nobel Prize site, and so on. Additionally, I agree with the list of positive features of e-commerce compiled by Jennifer Rowley in her book *E-Business: Principles and Practice* (Rowley, 2002), although I am unsure every generation and those living in extreme poverty will access the web:

- availability
  - 24-7 and immediate access
- ubiquity
  - it can be assumed that all organisations and customers will soon have internet access
- global
  - leaving delivery to one side, our mental map of near and far will radically change
- local
  - internet is also a good medium to reinforce local and physical presence and local person-to-person business relationships
- digitisation
  - business will increasingly be happening in information space. This leads to:
    - convergence of telecommunications, broadcasting and other information industries
    - different economic laws operating, with increasing rather than decreasing returns to scale
- multimedia
  - provides new opportunities for information provision during buying and selling and provides new opportunities in consultancy, design and entertainment

- interactivity
  - an opportunity to improve customer service at an affordable price
- one-to-one
  - using data processing and customer profiling, one-to-one marketing is a natural consequence of doing business on the internet. It may also be a necessity to overcome the anonymity of internet business relationships
- network effects and network externalities
  - low cost and fast growth in the number of relationships enable business models that require a significant number of parties on the network and whose benefits increase faster with a growing number of parties and/or network externalities
- integration
- the value of combined information across steps of the value chain is more than the sum of its parts. The internet now provides at least part of the technology for value-chain functional and information integration. Advanced electronic commerce companies show how to exploit the added value.

NOTES

1. http://www.queenonline.com/home.
2. http://www.imd.ch.
3. http://www.harley-davidson.com/wcm/Content/Pages/H-D_History/history_1900s.jsp?locale=en_US.
4. http://www.disneyinternationalprograms.com.
5. http://www.businessweek.com.
6. http://money.cnn.com/magazines/fortune/bestcompanies/2010/.
7. http://www.nytimes.com.
8. http://www.bell-labs.com/user/apenzias/profile1.html.
9. http://www.digitimes.com/topic/computex_taipei_2009/a00127.html.

REFERENCES

Dimancescu, D., and K. Dwenger. 1996. *World Class New Product Development: Benchmarking Best Practices of Agile Manufacturers*. New York: American Management Association. ISBN-10 0-814-40311-5.

Dussauge, P., S. Hart, and B. Ramanantsoa. 1992. *Strategic Technology Management.* Hoboken, NJ: John Wiley and Sons. ISBN-10 0-471-93418-6.

Emmott, W. 1992. *Japan's Global Reach.* London: Random House Business Books. ISBN-10 0-712-64928-X.

Matheson, D., and J. Matheson. 1998. *The Smart Organization.* Boston: Harvard Business School Press. ISBN-10 0-875-84765-X.

Mead, R. 2005. *International Management: Cross-Cultural Dimensions.* Hoboken, NJ: Blackwell Publishing. ISBN-10 0-631-23177-3.

Peters, T. 2003. *Re-Imagine! Business Excellence in a Disruptive Age.* London: Dorling Kindersley Limited. ISBN-10 1-405-30049-3.

Rousell, P. A., K. N. Saad, and T. J. Erickson. 1991. *Third Generation R&D.* Boston: Harvard Business School Press. ISBN-10 0-875-84252-6.

Rowley, J. 2002. *E-Business: Principles and Practice.* London: Palgrave. ISBN-10 0-333-94914-5.

Schneider, G. P., ed. 2007. *Electronic Commerce.* Independence, KY: CengageBrain. ISBN-13: 978-1418837037.

10

The "No-Change" Scenario and Possible Future Changes

Who in the world wants to come in second? No really, it's a serious point. You see, if all we do in business is to deploy lean thinking concepts in our companies, all that we will ever be able to do is try to catch up with companies such as the Toyota Motor Company. Toyota invented lean thinking and has refined and embedded the concepts into its organization and that of its suppliers so that its whole supply chain is now singing to the same tune. Most—not all, I admit—Western companies have waste in every sphere of activity; it may be small, but it will be there. So we need to eliminate waste by using the proven lean thinking concepts; but then what?

In general, when we have implemented lean thinking, we in the West will be where Toyota and others were in the 1980s. It seems to me that we should look to the future for what to do for those companies that are well down the road of implementing lean thinking. We've seen from the last three chapters that there are plenty of other proven concepts from other industries, such as process intensification from the world of chemical engineering and electronic data interchange from the world of e-commerce. Of course, some companies are just starting down the lean thinking route, and they need to implement the concepts before they can think beyond their current activities.

There is also another issue. Fewer companies in the world of e-business have the same hierarchical structure as the more traditional industries. Indeed, web-based companies often play by their own rules. Traditional companies are now becoming hybrid organizations with elements of web and traditional business practices. Best practices as exemplified by lean thinking may not necessarily "fit" into newer business models.

There is yet another problem. As we have also seen, the implementation of lean thinking is now embedded into the fabric of continuous improvement activities. Womack, Jones, and others have done some outstanding work[1] and should be justly proud of their achievements. We now have highly skilled change agents in the form of greenbelts, blackbelts, and master blackbelts who have the necessary skills to move from one project or organization to another and make significant progress over short timescales. These individuals have been trained to a high level and are skilled in what they do.

Suggesting that we need to look for the next quality improvement initiative beyond lean thinking might be counterproductive unless the "belts" are on board. The problem is that the whole lean thinking "structure," for want of a better word, is a loose association of companies and institutes with no central hub. In other words, if you wanted to update the concepts taught in "belt" training, how would it be done and rolled out? Furthermore, how would you update the knowledge base and skill sets of the current "belts" who are out in the real world doing outstanding work? Of course, there is yet another question, and that is "Why not leave the lean thinking folks to do their own thing and start a new quality initiative?"

Arguably the last point is easier to deal with than some of the others. All too often in industry, we have had various "initiatives" that have seemed to be the flavor of the month. Total quality control (TQC), Six Sigma, quality circles, and lean thinking for certain were all rolled out in the company where I worked. Unfortunately, the first few quality implementations did not have the chance to "bed in" before the next one was rolled out. Although they each tackle some of the same issues in different ways, each has its own and unique contribution to make. None was truly embedded in the culture.

Unfortunately, some of the communities within larger organizations do not necessarily see each system as something that can be embedded into their own part of the organization because they know that another initiative will come down from the head office before too long. One of the big advantages from the lean thinking approach is that there is one system to implement with time and resources continually devoted to the activity. In the past, each new CEO brought his or her own ideas when moving posts. The implementations of lean thinking are such that they are being applied to many industries, and so there is less churn when a new CEO takes the reins.

For the first time that I can remember, we have one system to consider with highly resourceful and trained personnel undertaking the implementation. This infrastructure should be nurtured and encouraged. Of course, the rest of the world has kept moving, as exemplified by the technology and new business processes identified in Chapters 7–9. Whether you agree with the topics covered in these chapters, or indeed have your own favorites that have not been included, there *are* proven new ideas out in the real world that are saving money on a daily basis. These new approaches coupled with the concepts of lean thinking may provide the business edge that will help Western industries to overtake the Eastern corporations.

Setting up a separate quality system that relies on new concepts unproven in some industry sectors will go nowhere. The controlled and structure improvement to an existing process (i.e., lean thinking) stands a better chance of success.

10.1 POSSIBLE CONTENDERS

There are many different processes designed and in use to assess the potential for research projects to enter into the development phase. Additionally, there are many *phases and gates processes* that will take development projects and ensure that the ideas can be considered for manufacturing. Many of these phases and gates processes are complex, having many criteria needing to be considered for each phase and gate. They are often bespoke to the industry for which they were developed partly as a result of the potential products under assessment.

Nevertheless, it would be useful to compare some of the concepts that have been presented in Chapters 7–9. I have decided to use the process that was developed by some workers at NASA in 1995 and published by John C. Mankins as a white paper on 6 April 1995.[2] Mankins worked in the Advanced Concepts Office at the Office of Space Access and Technology. The system he presented used *Technology Readiness Levels* (TRLs). Mankins described the process thus:

> Technology Readiness Levels (TRLs) are a systematic metric/measurement system that supports assessments of the maturity of a particular technology and the consistent comparison of maturity between different types of

technology. The TRL approach has been used on-and-off in NASA space technology planning for many years and was recently incorporated in the NASA Management Instruction (NMI 7100) addressing integrated technology planning at NASA.

There are nine TRL levels; TRL 1 is the lowest level, and TRL 9 is "flight proven." I have reworded some of the TRL definitions for brevity. It is worth looking at all nine levels of the TRLs because this highlights the issues of, and potential problems with, integrating other "elements" into lean thinking.

TRL 1. Basic principles observed and reported.
   Scientific research begins to be translated into applied research and development.
TRL 2. Technology concept and/or application formulated.
   Practical applications of those characteristics can be "invented" or identified.
TRL 3. Analytical and experimental critical function and/or characteristic proof of concept.
   Active research and development (R&D) is initiated. This must include both analytical studies to set the technology into an appropriate context and laboratory-based studies to physically validate that the analytical predictions are correct.
TRL 4. Component and/or breadboard validation in laboratory environment.
   Following successful "proof-of-concept" work, basic technological elements must be integrated to establish that the "pieces" will work together to achieve concept-enabling levels of performance for a component and/or breadboard.
TRL 5. Component and/or breadboard validation in relevant environment.
   At this level of maturity, the fidelity of the component and/or breadboard being tested has to increase significantly. The basic technological elements must be integrated with reasonably realistic supporting elements so that the total applications (at the component level, subsystem level, or system level) can be tested in a "simulated" or somewhat realistic environment.

TRL 6.   System or subsystem model or prototype demonstration in a relevant environment.

At this point, the maturation step is driven more by assuring management confidence than by R&D requirements. The demonstration might represent an actual system application, or it might only be similar to the planned application, but use the same technologies. At this level, several-to-many new technologies might be integrated into the demonstration.

TRL 7.   System prototype demonstration in a real environment.

The driving purposes for achieving this level of maturity are to assure system engineering and development management confidence (more than for purposes of technology R&D). Therefore, the demonstration must be of a prototype of that application.

TRL 8.   Actual system completed and "flight qualified" through test and demonstration (ground or space).

By definition, all technologies being applied in actual systems go through TRL 8. In almost all cases, this level is the end of true "system development" for most technology elements.

TRL 9.   Actual system "flight proven" through successful mission operations.

In almost all cases, the end of last "bug-fixing" aspects of true "system development." For example, small fixes or changes to address problems found following launch.

We'll have a look at my possible contenders for an expanded lean thinking model in a moment. Let us just consider some of the following issues:

- Each of the potential new "modules" has a proven track record in its own field.
- The potential new "modules" have different timescales for delivery.
- The potential new "modules" will need to be integrated with or complementary to existing lean thinking principles; otherwise, why incorporate into lean thinking?
- Why would a manager want to implement these new modules?

The last point is simple to answer. There will be senior managers who are looking at how they are going to increase market share, or reduce their company's environmental impact. Carbon credits can be traded, and some companies will plant trees in one part of the world so that they can pollute their own area. On the other hand, there are now newer companies, such as Better Place,[3] that have been set up to run with environmentally friendly goals built into their business strategies. In between these two extremes, there are firms that will look for technologies to speed up their transition to a "greener factory."

Yet a further motivation might be the increasing reliance on e-business and e-commerce. The lean thinking concepts of *kanbans* and painting lines on floors are great, cheap, and easy technology to implement. Unfortunately, these concepts in their current form do not lend themselves to e-business or e-commerce. The globalization of companies is leading to a greater reliance on virtual team meetings which require a solution for team coherence that relies on more modern concepts.

These and other business drivers will gently push, or in some cases force, senior managers to consider how to make the leap from lean thinking to the next level. Let's review the potential new modules from Chapters 7–9. Please bear in mind that the following list is just my suggestion. There will undoubtedly be other potential new modules to consider. Even if I have managed to capture all of the current "non–lean thinking" best practices, there will be others with time, so this list is dynamic.

10.1.1 Potential New Modules

People
   Interpersonal relationships: understanding self and others
   Teamship: enhanced levels of cooperation within a team
Process
   Intensification: the green agenda
   Verification: predictable processes
E-commerce
   Electronic data interchange: automatic supply chain ordering (*kanban* equivalent)

These are my top five candidates for inclusion in some form or another into lean thinking. Earlier in this chapter, we discussed the part Womack and Jones have played in making the Toyota Production System toolkit available to the rest of the world through lean thinking. James P. Womack is now chairman (and founder) of the Lean Enterprise Institute.[4] The institute posts the following on its website:

> The Institute conducts research activities in a wide range of industries to create a toolkit of methods for implementing lean thinking and the necessary leadership behaviours. The Institute also sponsors educational meetings, workshops, senior management seminars, and conferences through the year and helps people to apply lean thinking in manufacturing and entirely new applications such as healthcare, retail, air travel, and distribution.

Joseph Juran, who was mentioned in Chapter 1 (and died quite recently at the age of 103), founded the Juran Institute. The Juran Institute posts the following on its website:[5]

> The Juran Management System (JMS) is a comprehensive business management system that incorporates lessons learned from over 50 years of research and study by Dr. Joseph M. Juran and the Juran Institute. It is a system that began at Toyota in the 1950s and has continued to evolve over many decades. The JMS builds upon management principles as espoused in lectures, books, consulting, and most importantly, client results.
>
> The heart of the JMS is called the 'Juran Trilogy': Planning, Controlling and Improving the quality of products and processes. We believe that sustainable performance breakthroughs are achievable by following the principles of the Trilogy. Understanding the Trilogy is the first step in understanding the JMS. There are many other people and institutes which would be able to add to my list of potential new lean thinking modules, or indeed provide ideas of their own. From wherever the ideas originate, there may well be the need to find a mechanism for their inclusion as core parts of lean thinking or as additional training post blackbelt certification. If that were to happen, we could continue Mikel Harry's martial arts analogy (see Chapter 6), in which he associated different colors with belts to signify levels of knowledge and expertise. So we might have a first Dan skilled in additional people skills, a second Dan skilled in the new process methods, and a third Dan for e-commerce skills. The serious point here is that if these modules were to become accepted within industry, there will need to be descriptors so that job adverts can reflect the skills needed for an implementation, and those people who obtain the necessary skills have an identity within the community in which they serve.

So we have our potential new lean thinking modules, and we have the Technology Readiness Levels as an indicator of production readiness. Each potential new concept for adoption by lean thinking methodology will score a TRL level of 9 as they are all used in the commercial world in at least one industry.

The people skills modules have been taught on the MSc in business psychology within the Business Psychology Centre of the University of Westminster for at least the last five years. Graduates from these year groups have obtained employment in a variety of countries and industry sectors. These people skills concepts have therefore been deployed in a variety of companies. I would therefore suggest that these two new potential modules would score a TRL level for world use as TRL9 in their own industries.

Process intensification (PI) has been used to an ever-greater extent in the British chemical engineering industry, most notably ICI (now part of AkzoNobel). The concepts would need to be adapted for worldwide adoption and to any industry, and so PI would fall within the TRL levels of 4–6 depending on the application.

Process verification techniques are varied. Some of the large *human-machine interface* (HMI) software already contains modules for process verification. This module would therefore score a TRL level of 9.

Electronic data interchange (EDI) is a vital part of airplane manufacture at Boeing. It is a proven concept within the aero industry. As software links between disparate systems become easier to use, this technology will develop. I would suggest that this TRL level for a more general use would be in the 6–7 range. The point of this exercise is to demonstrate that these techniques *are* used each and every day in one or more industry sectors. Any organization wishing to apply these concepts for its industry would have a place to start.

The next obvious questions are "Why should I implement these techniques?" And "What's in it for me?" It is often difficult to financially quantify the benefits of increased awareness of people interactions. There are many companies that have implemented the Insights package. BT, formerly British Telecoms, has trained more than four thousand of its staff in its techniques and its particular take on Jungian psychology. Katrina Head, who is head of Leadership and People Development at BT Wholesale, posted the following comment on the Insights website:[6]

We use Insights' methodologies extensively throughout BT for understanding self, understanding others and learning how to adapt and connect more easily with everyone. We use the methodologies for team creation and team building to help managers understand their people more easily and to develop rewarding and productive relationships. The Insights Discovery language has swept throughout our organisation because it is fun to learn, inspirational to experience, easy to remember, effective and practical in application. This means it gets used. It is our objective to fully integrate Insights Discovery into our culture.

Although I do not wish to act as an unofficial spokesman for the Insights Company, it offers several such comments on its website from a variety of companies. Indeed, I was sufficiently impressed with its product while I was visiting professor in the Business Psychology Centre that I undertook the accreditation program.

Similarly, Professor Benton's Business Psychology Model[7] has received high praise from a number of organizations. Indeed, prior to the 2008 Olympics in Athens, Dr. Benton was invited to Athens to help with an aspect of teamwork that was in danger of delaying the building works. I doubt that the organizers would have called in any experts had they felt that they could have achieved their deadlines without external help.

In many respects, the cost benefits of process intensification are better understood. Although it is difficult to compile a national or world cost benefit, the BHR group posts the following on its website:[8]

Originally developed for the bulk chemical industry, PI developments at BHR Group and elsewhere have more recently been focused on the higher added-value chemicals and pharmaceutical active ingredient sectors.

First and foremost, PI (as practised at BHR Group) is a business driven approach—the focus is always on what business benefits are targeted and might be achieved. To ensure this is achieved, BHR Group has developed structured methodologies for the application of PI.

10.1.2 Features of PI Solutions

- Move from batch to continuous processing.
- Use of intensive reactor technologies with high mixing and heat transfer rates (e.g., FlexReactor and HEX Reactors) in place of conventional stirred tanks.

- Multidisciplinary approach, which considers opportunities to improve the process technology and underlying chemistry at the same time.
- "Plug-and-play" process technology to provide flexibility in a multi-product environment.

10.1.3 Established PI Benefits

Typical examples of established benefits are as follows:

- Capital cost reduced by 60 percent
- Ninety-nine percent reduction in impurity levels resulting in significantly more valuable product
- Seventy percent–plus reduction in energy usage and hence substantial reduction in operating cost
- Ninety-three percent yield first time out—better than a fully optimized batch process
- 99.8 percent reduction in reactor volume for a potentially hazardous process, leading to inherently safe operation

Process verification implementations are also fairly easy to quantify. From about 1995 to 2003, I worked on and then led a team of programmers and implementers whose job it was to implement process verification technology into various facets of the manufacturing processes on the site where I worked. At one stage, we were collecting approximately 6 Gbytes of data per day using streamlined and automated software systems to capture both analog and digital data. The analog signals ranged from pump speeds and temperatures to pressures and vacuum data. The digital in–out signals gave plant status indicators which were useful for fault diagnostics.

Finding a fault in one of the parameters using manual means was virtually impossible. We also installed software robots whose design allowed for data analysis of the results from which they populated relational databases. Wherever possible, aims, limits, and corrective action guidelines were written, and in some cases alarm limits created. The whole process was automated as much as possible.

These data were used in several ways. Should a fault occur that stopped the process, our systems allowed the engineers to more quickly diagnose faults and return the complex processes to manufacturing readiness. That became the first use of the data by shift engineers. The data were of much greater use to the day engineers, who could plan their maintenance days with the longer term trends in mind.

One of the problems with using one manufacturing process for producing eighty or so products is that some of the small runners use the process only for relatively short periods of time and perhaps only once or twice per year. The collected and analyzed data on their own did not produce sufficient data points for the small runners to correctly interpret these data for long-term process drift. However, our process was designed such that the previous data from as many older manufacturing runs as required could also be included in the analysis. This afforded longer term drift information previously denied to the process engineer.

Such trends could be used for longer term process evaluation, adding another useful benefit to the verification system. Such a complex system is difficult to implement and can be difficult to maintain, but is incredibly powerful in helping to do the following:

- Decrease mean time to recover (MTTR)
- Increase mean time between failure (MTBF)

These two measures can be quantified. The worldwide benefits of this system to my organization ran into tens if not hundreds of millions of dollars, more than a tenfold return on the investment.

Of course, I have described a complex system requiring specialized programmers and implementers. Some simple steps can be implemented that can save time and therefore money without the need to invest in an IT solution. Such systems also return far more than their investment costs.

You may recall from Chapter 9 the Boeing example of the benefit of electronic data interchange (EDI). (In summary, Boeing had a shutdown that cost the company several weeks of lost production costing more than $1.5 billion.) The EDI implementation has cut Boeing's delivery time from 36 months to 10–12 months for a new aircraft. EDI has delivered impressive savings for Boeing and will do so for smaller companies as e-commerce transactions force the rate of change.

So our contenders have demonstrated financial benefits to the bottom line, and are proven technologies in their own right. As I have stated in this chapter, these are just some processes that are available for implementation now. There may be others of equal standing in the business world unknown to me. That's OK; I don't have a monopoly on potential processes deserving a wider exposure and implementation. The next question to ask is "How would these potentially new lean thinking modules be integrated into the core toolkit?"

10.2 NEW MODULES, OLD PROCESS

During the time that Taiichi Ohno worked at the Toyota Motor Company, I doubt very much that he considered how much would be made of the implementation of his ideas by other industries. I have sat in all-day workshops sometimes called *boot camps*, where trainers from my former company demonstrated the benefits of the Toyota Production System that they had rebranded to suit the culture of my former company. There is nothing wrong with rebranding, as I mentioned in Chapter 3.

One of the diagrams favored in these training sessions was a jigsaw. The lean thinking implementers were fond of suggesting that each of the lean thinking elements could be likened to a jigsaw piece. They suggested that only when all of the pieces were put together was the picture whole. In other words, each facet of the Toyota Production System (TPS) needed to be implemented before all of the benefits could be realized. I actually agree with them and would suggest that the lean thinking modules should not be compromised as we consider upgrading the concepts to include new ideas.

My rationale for this is "If it ain't broke, don't fix it." In other words, there is much to be gained from implementing lean thinking as it currently stands. In fact, there are many successful parts of TPS. Replacing them for other potentially useful newer concepts may, and indeed would, be counterproductive.

I would also caution against increasing the training by another week or two just to include the newer material. The classroom training and projects take months to complete. Increasing the length of training would increase the time to cash to such an extent that small companies may decide that the process is too costly. And yet to my mind introducing another quality initiative would not gain the necessary penetration for it to be successful, so how do I suggest these newer concepts be implemented?

My suggestion is that the newer material be taught to established practitioners (blackbelts, greenbelts, and management blackbelts). There are many companies whose experts have successfully implemented lean thinking and now need modern ideas to take their businesses further. One or all of the five contenders mentioned earlier will be the place to start. In particular, I would suggest manufacturing companies consider process intensification methodology as these techniques will increase their "green footprint." This might take the form of reduced physical footprint, lower water or energy consumption, or increased operating efficiency.

As I mentioned earlier, all too often in the past, the various elements of lean thinking have been likened to jigsaws or even houses. These new ideas should be seen as modules that need to be grafted onto the parent, as a branch can be grafted onto a tree. To carry on with the analogy, the mainstay of the quality thrust should remain lean thinking, represented by the tree trunk. The newer modules should be seen as branches grafted onto the main trunk. Hopefully, Figure 10.1 better demonstrates this concept.

Just as with any tree, the number of branches is not fixed. New modules could be added as and when a proven technology is worthy of inclusion. In this particular manifestation, there are five main branches. With time,

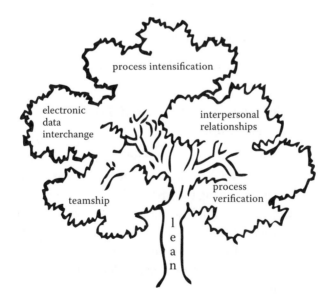

FIGURE 10.1
The lean thinking tree of techniques.

there could be a sixth, seventh, and eighth. The point is that we should not restrict ourselves to a fixed number. We should add and modify the "tree" to cope with changing circumstances. In other words, it should take on a life of its own and in a way become *organic*. This will allow for evolution rather than a fixed silo mentality. If you pardon the pun, we should take a leaf from nature's book and go with the flow.

It's obvious, however, that the new modules have to be grafted onto the original lean thinking concepts such that there is a seamless transition. They have to feel as if they "belong"; otherwise, it is a pointless exercise. As we all know, there are some best practices that have nothing to do with waste reduction. Unrelated ideas would distract from the purpose of lean thinking.

Some pointers for including a new module to the "techniques tree" might be the following:

- Does the new idea reduce waste?
- Is the new idea an extension of an original lean concept?
- Will the new idea add value or cause confusion?
- Will the new technique complement or extend lean thinking?

Let's have a look at our five modules using the above "guidelines."

10.2.1 Does the New Idea Reduce Waste?

All five proposed modules do indeed reduce waste in one form or another.

10.2.2 Is the New Idea an Extension of an Original Lean Concept?

- Both interpersonal relationships and teamship modules enhance levels of cooperation within a team. They extend or complement the "soft skills" taught during the first week of most blackbelt courses.
- Process intensification implementations have been shown to be closely linked to just-in-time manufacturing and also to inventory reduction.
- Process verification is used to prevent defects from being produced by the process and is closely linked to *jidoka* (stopping the process if faults are detected).
- Electronic data interchange is closely linked to inventory reduction and just-in-time parts supply.

10.2.3 Will the New Idea Add Value or Cause Confusion?

As we have seen from some information presented in this chapter, there are sound business paybacks to implementing all five of these modules.

10.3 IMPLEMENTATION

A web search for *lean thinking, lean manufacturing,* and *Toyota Production System* yielded over 3.6 million hits in December 2009. That's a lot! Accepting that some hits could be duplicates, there is still a lot of interest in lean thinking concepts and implementation. That's both a good thing and actually also a bad thing.

I mentioned earlier that there does not appear to be an overall accreditation system for organizations that want to teach and consult on lean concepts. I therefore have a slight problem when organizations advertise their training as being certified! I always ask myself the question of how they achieved their certification status if there is no international regulatory body.

Let's have a look at the good side first. With no formal regulation, there should be no problem in teaching new modules grafted onto the formal lean thinking. This would be the case where training is provided to blackbelts who are already accredited and who have been practitioner or even master blackbelts for some time. Such courses could take place in institutes where there are leading practitioners of the module content. In this way, the subject is taught by the experts.

As this additional training is modular, there is no reason why each candidate blackbelt should attend all of the five modules. He or she could simply attend the module pertaining to current needs. Additionally, there is no reason for one institute to teach all five modules. Blackbelts and master blackbelts may wish to specialize in one or more of the newer techniques. This should not be a problem as we already have bachelor, master, and doctorate degrees. The former course teaches the general subject, and the postgraduate degrees offer a more in-depth training in particular aspects of the bachelor degree. Appropriate training could also be offered in this way for greenbelts.

I appreciate that this concept would lead to blackbelts who have undertaken training in one or more extra modules that would need to be reflected in their title or CV. I have already suggested using the *Dan* nomenclature (although this was very much tongue in cheek!). As time goes by, more modules could be added to the advanced training, forcing this issue to be resolved.

However, adding new modules is where the fun starts. With no regulatory body, a host of blackbelt add-on courses could spring up all over the globe. There may come a time where there is chaos. Both blackbelts and master blackbelts will need to be careful when applying to these courses. This problem will be overcome, at least in part, if the further training takes place, as I suggest above, only at the centers of excellence for the new modules. Another downside is the numbers of blackbelts worldwide. There may be a need to create an accreditation process for the advanced-skills "belts." My reasoning here is that the centers of excellence may not have the capacity to complete the necessary training for the numbers wanting to update their skills.

An accreditation process would allow the centers of excellence to license other institutions. This would ensure that only state-of-the-art proven techniques enter the new syllabus. Well-grafted advanced modules will take advantage of the excellent work of Womack, Jones, and others as well as that of the "belts" themselves.[9] Properly implemented modules should be grafted onto lean thinking methods, producing a quality improvement and waste reduction juggernaut. Hopefully the knowledge and momentum will be such that the combined and updated lean thinking concepts will help industry to surpass Eastern techniques.

So there we have it.

Let's just finish with some wise words. Sir Gerry Robinson is a British business expert. He has been chairman of Allied Domecq PLC, Granada, the Arts Council of England, British Sky Broadcasting Group, and ITN. He was also appointed managing director of Grand Metropolitan's troubled International Services business and went on to become chief executive of the whole Contract Services Division. He has been heralded as one of Britain's best businessmen. Just recently he examined the car manufacturing industry on television, a program that was co-produced between the BBC and the Open University. I have paraphrased his conclusion. He suggested that we need to do the following:

- Move to greener products—this is essential.
- Concentrate on the manufacture of high-end and/or advanced products.
- Develop the skill set of our designers and engineers (and hang on to the intellectual property).

While there are no surprises, his conclusion fits in with some of the proposed advanced modules. Additionally, Bob Sutton, a professor of industrial engineering at Stanford University, wrote a book called *Weird Ideas That Worked: 11½ Practices for Promoting, Managing and Sustaining Innovation* (Sutton, 2001). Eleventh on the list is of interest here, but just for completeness I have reproduced the whole list below.

1. Hire slow learners (of the organizational code).
1½. Hire people who make you uncomfortable, even those you dislike.
2. Hire people you (probably) don't need.
3. Use job interviews to get ideas, not to screen candidates.
4. Encourage people to ignore and defy superiors and peers.
5. Find some happy people and get them to fight.
6. Reward success and failure, and punish inaction.
7. Decide to do something that will probably fail, then convince yourself and everyone else that success is certain.
8. Think of some ridiculous, impractical things to do, then do them.
9. Avoid, distract, and bore customers, critics, and anyone else who just wants to talk money.
10. Don't try to learn anything from people who seem to have solved the problems that you face.
11. Forget the past, particularly your company's success.

The last point is really useful to remember. The past does not guarantee the future, for there are many Fortune top 500 companies from previous years that are no longer on that list. Eastman Kodak and Polaroid have suffered severe financial constraints with the onset of digital photography. Their success in previous decades was not able to protect them.

No one can afford to be complacent and expect to command dominance in global markets, especially with e-commerce. This particular business outlet never sleeps as the web is available 24/7. There are always buyers looking for high-quality goods at reasonable prices. Coming second should not be a serious option. We should always be on the lookout for a financial edge.

Finally, I always thought the following phrase was originally said by Tom Peters. I contacted him and asked as I couldn't find it written in any of his books. His assistant replied that he didn't remember ever saying the phrase, although it sounded as if it was one of his sayings. It therefore remains anonymous.

That you must always worry about the competitor. If you have less than 100 percent of the market, *someone* likes the competitor's product better than yours. The competitor's 1 percent may become 2 percent, and then 3 percent. An overall loss of 0.25 percent might be a 20 percent loss in an important market niche.

NOTES

1. http://www.leanuk.org.
2. http://www.hq.nasa.gov/office/codeq/trl/trl.pdf.
3. http://www.betterplace.com.
4. http://www.lean.org/WhoWeAre/LeanPerson.cfm?LeanPersonId=1.
5. http://www.juran.com/HomeLeftNav/juran_mgt_system.aspx.
6. http://www.insights.com/index.aspx.
7. http://businesspsychologycentre.com.
8. http://www.bhrgroup.co.uk/pi/aboutpi.htm.
9. http://www.leanuk.org.

REFERENCE

Sutton, R. I. 2001. *Weird Ideas That Worked: 11½ Practices for Promoting, Managing and Sustaining Innovation.* New York: Free Press. ISBN-10: 0-743-21212-6, ISBN-13: 978-0743212120.

Index